Advanced Issues in Partial Least Squares Structural Equation Modeling

Second Edition

T0323304

Sara Miller McCune founded Sage Publishing in 1965 to support the dissemination of usable knowledge and educate a global community. Sage publishes more than 1,000 journals and over 600 new books each year, spanning a wide range of subject areas. Our growing selection of library products includes archives, data, case studies, and video. Sage remains majority owned by our founder and after her lifetime will become owned by a charitable trust that secures the company's continued independence.

Los Angeles | London | New Delhi | Singapore | Washington DC | Melbourne

Advanced Issues in Partial Least Squares Structural Equation Modeling

Second Edition

Joseph F. Hair, Jr.

University of South Alabama

Marko Sarstedt

Ludwig-Maximilians-University

Christian M. Ringle

Hamburg University of Technology

Siegfried P. Gudergan

James Cook University

S Sage

FOR INFORMATION:

2455 Teller Road
Thousand Oaks, California 91320
E-mail: order@sagepub.com

1 Oliver's Yard
55 City Road
London, EC1Y 1SP
United Kingdom

Unit No. 323-333, Third Floor, F-Block
International Trade Tower
Nehru Place, New Delhi – 110 019
India

18 Cross Street #10-10/11/12
China Square Central
Singapore 048423

Printed in the United States of America

ISBN 978-1-0718-6250-6

This book is printed on acid-free paper.

Acquisitions Editor: Leah Fargotstein

Product Associate: Emma Newsom

Production Editor: Jothi Karan

Copy Editor: Talia Greenberg

Typesetter: TNQ Technologies

Cover Designer: Candice Harman

Marketing Manager: Victoria Velasquez

23 24 25 26 27 10 9 8 7 6 5 4 3 2 1

BRIEF CONTENTS

Preface   xi

About the Authors   xvii

Chapter 1   An Overview of Recent and Emerging Developments
                 in PLS-SEM   1

Chapter 2   Higher-Order Constructs   33

Chapter 3   Advanced Modeling and Model Assessment   57

Chapter 4   Advanced Results Illustration   91

Chapter 5   Modeling Observed Heterogeneity   133

Chapter 6   Modeling Unobserved Heterogeneity   167

Glossary   201

References   211

Index   231

DETAILED CONTENTS

Preface   xi

About the Authors   xvii

Chapter 1   An Overview of Recent and Emerging Developments in PLS-SEM   **1**

   Chapter Preview   1

   Origins and Evolution of PLS-SEM   1

   Model Specification   5

   Model Estimation   9

     The Original PLS-SEM Algorithm   9

     The Weighted PLS-SEM Algorithm   11

     Consistent PLS-SEM   12

   Principles of PLS-SEM   14

     Philosophy of Measurement   14

     Parameter Estimation Accuracy   15

     Model Estimation Implications   19

   Organization of the Remaining Chapters   20

   Case Study Illustration   23

     Corporate Reputation Model   23

     PLS-SEM Software   26

     Setting Up the Model in SmartPLS   26

   Summary   29

   Review Questions   30

   Critical Thinking Questions   30

   Key Terms   31

   Suggested Readings   31

Chapter 2   Higher-Order Constructs   **33**

   Chapter Preview   33

   Higher-Order Constructs   33

     Terminology and Motivation   33

     Types of Higher-Order Constructs   36

Specifying Higher-Order Constructs   41

    Overview   41

    The Embedded Two-Stage Approach   42

    The Disjoint Two-Stage Approach   43

    Recommendations   43

Estimating Higher-Order Constructs   44

Validating Higher-Order Constructs   45

Rules of Thumb   47

Case Study Illustration   47

Summary   53

Review Questions   54

Critical Thinking Questions   54

Key Terms   55

Suggested Readings   55

Chapter 3   Advanced Modeling and Model Assessment   57

Chapter Preview   57

Nonlinear Relationships   58

    Modeling Quadratic Effects in PLS-SEM   61

    Evaluation of Nonlinear Effects   63

    Results Interpretation   65

    Case Study Illustration   66

Confirmatory Tetrad Analysis   71

    Method   71

    Case Study Illustration   79

Summary   86

Review Questions   87

Critical Thinking Questions   88

Key Terms   88

Suggested Readings   89

Chapter 4   Advanced Results Illustration   91

Chapter Preview   91

Necessary Conditions Analysis   92

    Purpose and Method   92

    Identifying Necessary Conditions   94

    Considerations When Running an NCA   97

    Results Interpretation   98

    Case Study Illustration   99

Importance-Performance Map Analysis   105
  Overview   105
  Systematic IPMA Execution   107
    Step 1: Requirements Check   107
    Step 2: Computation of the Performance Values   109
    Step 3: Computation of the Importance Values   112
    Step 4: Importance-Performance Map Creation   116
    Step 5: Extension of the IPMA to the Indicator Level   117
  Case Study Illustration   119

Summary   128

Review Questions   129

Critical Thinking Questions   130

Key Terms   130

Suggested Readings   130

Chapter 5   Modeling Observed Heterogeneity   133

Chapter Preview   133

Observed and Unobserved Heterogeneity   134

Testing for Measurement Model Invariance   136
  Step 1: Configural Invariance   138
  Step 2: Compositional Invariance   139
  Step 3: Equality of Composite Mean Values and Variances   142

Multigroup Analysis   144
  Parametric MGA   145
  Bootstrap MGA   146
  Permutation Test   148
  Comparing More Than Two Groups   149
  Recommendations   150

Case Study Illustration   152

Summary   163

Review Questions   164

Critical Thinking Questions   164

Key Terms   165

Suggested Readings   165

Chapter 6   Modeling Unobserved Heterogeneity   167

Chapter Preview   167
  Uncovering Unobserved Heterogeneity in PLS Path Models   168
  Step 1: Run the FIMIX-PLS Procedure   171

Step 2: Determine the Number of Segments   172
Step 3: Run the PLS-POS Procedure   175
Step 4: Explain the Latent Segment Structure   180
Step 5: Analyze group-specific models   182

Case Study Illustration   184
Step 1: Run the FIMIX-PLS Procedure   184
Step 2: Determine the Number of Segments   186
Step 3: Run the PLS-POS Procedure   188
Step 4: Explain the Latent Segment Structure   190
Step 5: Analyze group-specific models   193

Summary   197

Review Questions   198

Critical Thinking Questions   199

Key Terms   199

Suggested Readings   200

Glossary   **201**

References   **211**

Index   **231**

PREFACE

Since the publication of the first edition of *Advanced Issues in Partial Least Squares Structural Equation Modeling (PLS-SEM)* 2018, the field has undergone massive developments. PLS-SEM use has surged, not only in fields that popularized the method such as management information systems, marketing, and strategic management (e.g., Sarstedt, Hair, Pick, Liengaard, Radomir, & Ringle, 2022), but in various other disciplines such as agriculture, engineering, environmental sciences and ecology, geography, and psychology (Sarstedt, 2019). PLS-SEM's methodological features have vastly extended researchers' capacities to better understand the complex interrelationships that constitute the "black box" of a variety of attitudinal and behavioral theories as well as explain and predict unobservable phenomena (e.g., Petter, 2018; Petter & Hadavi, 2021; Russo & Stol, 2023; Sharma, Liengaard, Hair, Sarstedt, & Ringle, 2023). For example, PLS-SEM has been used to predict consumers' vaccination behavior (Nguyen, Tsai, Lin, & Hu, 2021) and travel intentions during the COVID-19 pandemic (Şengel et al., 2022) and understand the factors that drive online information avoidance (Sultana, Dhillon, & Oliveira, 2023) as well as social support perceptions during lockdown periods (Sitar-Taut, Mican, Frömbling, & Sarstedt, 2022). Many of these analyses rely on advanced modeling and model assessment routines, such as Zhou and Wang's (2022) evaluation of conditions that facilitated pro-environmental behavior during the COVID-19 pandemic. These authors combined PLS-SEM with the necessary condition analysis (NCA; Dul, 2016), which extends the standard results illustration by identifying potential must-have conditions (e.g., certain scores of antecedent constructs) that need to be present for producing a certain outcome (e.g., a certain degree of pro-environmental behavior). While the combination of PLS-SEM and NCA has only recently been proposed (Richter et al., 2020), other advanced analysis options are well established in the field. Examples include the use of higher-order constructs (Sarstedt et al., 2019) and the treatment of observed or unobserved heterogeneity by means of measurement invariance assessment, multigroup analysis, and latent class analysis (Sarstedt, Ringle, & Hair, 2017).

Developments extend beyond the introduction of new methods. As science progresses, so does our understanding of the strengths and weaknesses of certain methodological extensions and modeling options. For example, Klesel, Schuberth, Henseler, and Niehaves's (2022) systematic evaluation of multigroup analysis procedures has not only confirmed prior recommendations regarding the choice of the most appropriate approach (Sarstedt, Henseler, & Ringle, 2011), but also extended the knowledge on how to conduct multigroup analysis. Other methodological developments have had an effect on how researchers apply certain advanced methods. For example, findings in the field of model comparisons have had immediate implications for researchers working with nonlinear effect models as they can now draw on set routines for benchmarking their model specifications in terms of fit and predictive power (e.g., Sharma, Shmueli, Sarstedt, & Danks, 2018). Similarly, the proliferation of Shmueli et al.'s (2016) PLS$_{predict}$ procedure has had important implications for the way researchers specify higher-order constructs or evaluate the appropriateness of latent class analyses.

Although most of these advances are documented in extant literature, authors, reviewers, and editors may find keeping up with the latest developments daunting, leading to the relevant complementary methods and guidelines only being adopted slowly. With this issue in mind, the second edition of *Advanced Issues in Partial Least Square Structural Equation Modeling* sets out to explain advanced modeling, results evaluation, and illustration options in PLS-SEM for researchers who are relatively new to the field. Building on the previous contents, the new edition comes with various updates to reflect the state-of-research in the field as well as new methods. These include the following:

- Overview of most recent developments in PLS-SEM.

- Detailed coverage of specification types of higher-order constructs (e.g., embedded and disjoint two-stage approaches).

- New guidelines for estimating and validating higher-order constructs.

- More details on discriminant validity testing in reflectively specified higher-order models.

- Coverage of the NCA that extends the standard PLS-SEM results assessment by assuming a necessity logic.

- More details on the permutation procedure in multigroup analysis.

- Updates on recommendations regarding multigroup analysis.

- Further details on comparing more than two groups using multigroup analysis.

- More information on the performance of information criteria in FIMIX-PLS.

- Extended guidelines for choosing the number of segments.

The approach of this book is based on the authors' many years of conducting research and teaching, as well as the desire to communicate the PLS-SEM method to a broader audience. To accomplish this goal, we have limited the emphasis on equations, formulas, Greek symbols, and so forth that are typical of most books and articles. Instead, we explain in detail the basic fundamentals of PLS-SEM and provide rules of thumb that can be used as general guidelines for understanding and evaluating the results of applying the method. As a further effort to facilitate learning, we use a single case study throughout the book. The case is drawn from a published study on corporate reputation and we believe is general enough to be understood by many different areas of social science research, thus further facilitating comprehension of the method. Review and critical thinking questions are posed at the end of the chapters, key terms are defined to better understand the concepts, and suggested readings and extensive references are provided to enhance more advanced coverage of the topic. We also rely on a single software package (SmartPLS 4; https://www.smartpls.com) that can be used not only to complete the exercises in this book but also in the reader's own research.

Despite its pedagogical features, the book is advanced in that it builds upon readers' knowledge of the foundation of PLS-SEM. Hence, readers are advised to read the third edition of the *Primer on Partial Least Squares Structural Equation Modeling (PLS-SEM)* (Hair, Hult, Ringle, & Sarstedt, 2022), which provides an introduction to the method and is extremely popular among researchers and practitioners (Ringle, 2019). A workbook version for the statistical environment R that covers the key steps of a PLS-SEM analysis is also available free of charge:

Hair, Hult, Ringle, Sarstedt, Danks, and Ray (2022). Partial Least Squares Structural Equation Modeling (PLS-SEM) Using R. webpage at: https://link.springer.com/book/10.1007/978-3-030-80519-7.

We have also set up a webpage at https://www.pls-sem.net that includes various support materials to facilitate learning and applying the PLS-SEM method. Additionally, the PLS-SEM Academy (https://www.pls-sem-academy.com) offers video-based online courses covering the foundations and various extensions of the method. Exhibit 0.1 explains how owners of this book can obtain a discounted access to the courses offered by the PLS-SEM Academy.

EXHIBIT 0.1 ■ DISCOUNTED PLS-SEM ACADEMY ACCESS

The PLS-SEM Academy (https://www.pls-sem-academy.com) offers video-based online courses on the PLS-SEM method. The courses include contents such as:

- An introduction to PLS-SEM,
- Reflective and formative measurement model assessment,
- Structural model assessment, goodness of fit, and predictive model assessment,
- Advanced topics, like importance-performance-map analysis (IPMA), mediation, higher-order models, moderation, moderated-mediation, measurement invariance, multigroup analysis, necessary condition analysis (NCA), and nonlinear effects.

Besides several hours of online video material presented by world-wide renowned instructors, the PLS-SEM Academy provides comprehensive lecturing slides and annotated outputs from SmartPLS that illustrate all analyses step-by-step. Registered users can claim course certificates after successful completion of each section's exam.

The PLS-SEM Academy offers all owners of this book a 15% discount on the purchase of access to its course offerings. All you have to do is send a photo of yourself with the book in your hand, your name and address to the e-mail address support@pls-sem-academy.com. A short time later you will receive a 15% discount code that you can use on the website https://www.pls-sem-academy.com. We hope you enjoy perfecting your PLS-SEM skills with the help of these courses and wish you every success in obtaining the certificates.

We would like to acknowledge the many insights and suggestions provided by a number of our colleagues and students. Most notably, we thank Jan-Michael Becker (BI Norwegian Business School), Zakariya Belkhamza (Ahmed Bin Mohammed Military College), Diógenes de Souza Bido (Universidade Presbiteriana Mackenzie), Charla Brown (Troy University), Roger Calantone (Michigan State University), Fabio Cassia (University of Verona), Jamie Carlson (University of Newcastle), Gabriel Cepeda Carrión (University of Seville), G. Tomas M. Hult (Michigan State University), Jacky Cheah Jun Hwa (University of East Anglia), Nicholas Danks (Trinity College Dublin), Adamantios Diamantopoulos (University of Vienna), Markus Eberl (Temedica), George Franke (University of Alabama), Rasoul Gholamzade (Payame Noor University), Anne Gottfried (University of Texas, Arlington), Karl-Werner Hansmann (University of Hamburg), Sven Hauff (Helmut Schmidt University), Philip Holmes (Pensacola Christian College), Chris Hopkins (Auburn University), Lucas Hopkins (Florida State University), Heungsun Hwang (McGill University), Ida Rosnita Ismail (Universiti Kebangsaan Malaysia), Morten B. Jensen (Aarhus University), April Kemp (Southeastern Louisiana University), David Ketchen (Auburn University), Ned Kock (Texas A&M University), Marcel Lichters (OVGU Magdeburg), Benjamin Liengaard (Aarhus University), Yide Liu (Macau University of Science and Technology), Francesca Magno (University of Bergamo), Vasilica Maria Margalina (Central University of Catalonia), Lucy Matthews (Middle Tennessee State University), Mumtaz Ali Memon (NUST Business School), Adam Merkle (University of South Alabama), Ovidiu I. Moisescu (Babes-Bolyai University), Zach Moore (University of Louisiana at Monroe), Forrest V. Morgeson (Michigan State University), Arthur Money (Henley Business School), Christian Nitzl (Universität der Bundeswehr München), Torsten Pieper (University of North Carolina), Dorian Proksch (University of Twente), Lacramioara Radomir (Babes-Bolyai University), Arun Rai (Georgia State University), Sascha Raithel (Freie Universität Berlin), S. Mostafa Rasoolimanesh (Taylor's University), Soumya Ray (National Tsing Hua University), Nicole Richter (University of Southern Denmark), Edward E. Rigdon (Georgia State University), Jeff Risher (Southeastern Oklahoma University), José Luis Roldán (University of Seville), Phillip Samouel (University of Kingston), Francesco Scafarto (University of Rome "Tor Vergata"), Rainer Schlittgen (University of Hamburg), Manfred Schwaiger (Ludwig-Maxmillians University Munich), Pratyush N. Sharma (University of Alabama), Wen-Lung Shiau (Zhejiang University of Technology), Galit Shmueli (National Tsing Hua University), Donna Smith (Ryerson University), Detmar W. Straub (Georgia State University), Hiram Ting (UCSI University), Ron Tsang (Agnes Scott College), Ramayah Thurasamy (Universiti Sains Malaysia), Huiwen Wang (Beihang University), Sven Wende (SmartPLS GmbH), Ronald Tsang (Agnes Scott College), and Anita Whiting (Clayton State University) for their helpful remarks.

We would like to acknowledge the many insights and suggestions provided by the reviewers for this edition: Saurabh Gupta (Kennesaw State University), Anantharam Narayanan (University of Cincinnati), and Jason Xiong (Appalachian State University). Also, we thank the teams at Hamburg University of Technology and Ludwig-Maximilians University Munich—namely, Susanne Adler, Janna Ehrlich, Lena Frömbling, Benjamin Maas, Carola Neumann, Lea Rau, Tobias Regensburger, and Lisa Schreiber—for their kind support. In addition, at SAGE we thank Leah Fargotstein for her support and great work.

We hope this book will expand knowledge of the capabilities and benefits of PLS-SEM to a much broader group of researchers and practitioners. Finally, if you have any remarks, suggestions, or ideas to improve this book, please get in touch with us. We appreciate any feedback on the book's concept and contents!

Joseph F. Hair, Jr., University of South Alabama

Marko Sarstedt, Ludwig-Maximilians-University, Munich, Germany,
and Babeș-Bolyai University, Romania

Christian M. Ringle, Hamburg University of Technology, Germany

Siegfried P. Gudergan, James Cook University, Australia,
and Aalto University, Finland,
and Vienna University of Economics and Business (WU Wien)

ABOUT THE AUTHORS

Joseph F. Hair, Jr. is professor of marketing, PhD director, and the Cleverdon Chair of Business in the Mitchell College of Business, University of South Alabama. He previously held the Copeland Endowed Chair of Entrepreneurship and was director of the Entrepreneurship Institute, Ourso College of Business Administration, Louisiana State University. He has authored over 95 books, including *Multivariate Data Analysis* (8th ed., 2019; cited 170,000+ times), *MKTG* (13th ed., 2019), *Essentials of Business Research Methods* (5th ed., 2023), and *Essentials of Marketing Research* (6th ed., 2023). Professor Hair is the most highly cited scholar in PLS-SEM and marketing, with 340,000+ citations (Google Scholar, 2023). He also has published numerous articles in scholarly journals and was recognized as the Academy of Marketing Science Marketing Educator of the year. A popular guest speaker, Professor Hair often presents seminars on research techniques, multivariate data analysis, and marketing issues for organizations in Europe, Australia, China, India, and South America.

Marko Sarstedt is professor of marketing at the Ludwig Maximilians University of Munich (Germany) and an adjunct research professor at Babeş-Bolyai University at Cluj-Napoca (Romania). His main research interest is the advancement of research methods to further the understanding of consumer behavior. His research has been published in *Nature Human Behaviour, Journal of Marketing Research, Journal of the Academy of Marketing Science, Multivariate Behavioral Research, Organizational Research Methods, MIS Quarterly, British Journal of Mathematical and Statistical Psychology*, and *Psychometrika*, among others. His research ranks among the most frequently cited in the social sciences, with more than 200,000 citations according to Google Scholar. Marko has won numerous best paper and citation awards, including five Emerald Citations of Excellence awards and two AMS William R. Darden Awards. Marko has been repeatedly named a member of Clarivate Analytics' Highly Cited Researchers List. In March 2022, he was awarded an honorary doctorate from Babeş-Bolyai University for his research achievements and contributions to international exchange.

Christian M. Ringle is professor of management and decision sciences at the Hamburg University of Technology (Germany). His research focuses on management and marketing topics, method development, business analytics, machine learning, and the application of business research methods to decision making. His contributions have been published in journals such as *International Journal of Research in Marketing, Information Systems Research, Journal of the Academy of Marketing Science, MIS Quarterly, Organizational Research Methods*, and *The International Journal of Human*

Resource Management. Since 2018, he has been named member of Clarivate Analytics' Highly Cited Researchers List. In 2014, Ringle cofounded SmartPLS (https://www .smartpls.com), a software tool with a graphical user interface for the application of the partial least squares structural equation modeling (PLS-SEM) method. Besides supporting consultancies and international corporations, he regularly teaches doctoral seminars on business analytics and multivariate statistics, the PLS-SEM method, and the use of SmartPLS worldwide. More information about Professor Ringle can be found at https://www.tuhh.de/hrmo/team/prof-dr-c-m-ringle.html.

Siegfried P. Gudergan is professor of strategy within James Cook University in Australia. He also is a visiting distinguished professor at Aalto University in Finland, visiting professor at Vienna University of Economics and Business (WU Wien) in Austria, and emeritus professor at the University of Waikato in New Zealand. He holds a PhD in management from the Australian Graduate School of Management that was awarded by both the University of Sydney and the University of New South Wales in Australia, and a degree in business and economics from RWTH-Aachen University in Germany. His research has been published in leading management, strategy, and marketing journals, and is recognized internationally. Some of his PhD students have won awards from the Academy of Management and Strategic Management Society. He has various board and professional roles, and has consulted or worked with numerous organizations, including start-ups and large multinational companies.

1 AN OVERVIEW OF RECENT AND EMERGING DEVELOPMENTS IN PLS-SEM

LEARNING OUTCOMES

1. Understand the origins and evolution of PLS-SEM.

2. Comprehend model specification in a PLS-SEM framework.

3. Describe the PLS-SEM algorithm's basic functioning principles.

4. Understand PLS-SEM's key characteristics vis-à-vis CB-SEM.

CHAPTER PREVIEW

Along with the recent surge in applications of partial least squares structural equation modeling (PLS-SEM), methodological research has prompted numerous extensions of the original method that vastly increase its scope. In this chapter, we first provide an overview of the origins and evolution of PLS-SEM. This foundation will enable us to better understand why the method was slow to be adopted in the beginning, but has been increasingly applied in recent years across many social science disciplines, particularly in the various fields of business administration. We then discuss different aspects related to the specification of measurement and structural models, followed by a brief introduction of the PLS-SEM algorithm and selected extensions. Several considerations, which have their roots in the characteristics of the method, are important when applying PLS-SEM. We therefore discuss important characteristics of the PLS-SEM method that relate to the underlying measurement philosophy and the implications that arise from the way the algorithm estimates the model parameters. The chapter concludes with the introduction of a case study used throughout the remainder of this book.

ORIGINS AND EVOLUTION OF PLS-SEM

The precursors to the PLS-SEM method were two iterative procedures that used least squares estimation to develop solutions for single and multicomponent models, and also for the method of canonical correlation (Wold, 1966). Further development of these procedures by Herman Wold led to the nonlinear iterative partial least squares (NIPALS) algorithm (Wold, 1973). A subsequent generalized version of the PLS-SEM algorithm focused on establishing and including latent variables in path models (Lohmöller, 1989, Chapter 2; Wold, 1980, 1982, 1985).

Several PLS methods evolved from Wold's generalized least squares algorithm (Mateos-Aparicio, 2011). One method is **principal components regression** (Hotelling, 1957; Jolliffe, 1982; Kendall, 1957; McCallum 1970), which performs a principal component analysis on the independent variables in which the model components are used as explanatory variables for a single dependent variable. However, principal components regression focuses on reducing the dimensionality of the independent variables only without considering the relationship between the independent and dependent variables.

Another method is the **partial least squares regression (PLS-R)**, which was originally designed to reduce the issue of collinearity in regression models (Abdi, 2010; Kiers & Smilde, 2007; Wold, Ruhe, Wold, & Dunn, 1984). PLS-R focuses on the dimension reduction of the independent variables in a regression model intending to remove collinearity from the predictor variables. By doing so, the method optimizes the variance extracted from the independent variables while simultaneously maximizing the variance explained in the dependent variables. More precisely, PLS-R relies on a principal component analysis that extracts linear composites of the independent variables and their respective scores. Its aim is to reduce the dimensionality of the independent variables, while taking the relationship between the independent and dependent variables into consideration. As a result, PLS-R enables researchers to estimate models with more independent variables than observations in the dataset (Valencia & Diaz-Llanos, 2003).

Interestingly, PLS-R was not developed by Herman Wold but by his son, Svante Wold (e.g., Wold, Sjöström, & Eriksson, 2001), who worked in the field of analytical chemistry, known today as chemometrics. Together with Harald Martens, he adapted NIPALS to analyze chemical data. In addition to addressing the problem of multicollinearity in multiple regression models, their method solved the problem that arises when the number of variables is larger than the number of respondents (Martens, Martens, & Wold, 1983).

A third method that emerged from Wold's (1980) generalized PLS algorithm for estimating relationships between constructs and their indicators as well as between constructs (Lohmöller, 1989; Wold, 1982), also referred to as **partial least squares path modeling** (e.g., Esposito Vinzi, Trinchera, & Amato, 2010; Tenenhaus, Esposito Vinzi, Chatelin, & Lauro, 2005), the PLS approach to structural equation modeling (e.g., Chin, 1998), and **partial least squares structural equation modeling (PLS-SEM**; e.g., Hair, Ringle, & Sarstedt, 2011; Sarstedt, Hair, & Ringle, 2023). PLS-SEM determines the parameters of a set of equations in a path model by combining principal components analysis to assess the measurement models with path analysis to estimate the relationships between constructs (i.e., latent variables). Wold (1982, 1985) proposed his "soft modeling basic design" underlying PLS-SEM as an alternative to Jöreskog's (1973) **covariance-based structural equation modeling (CB-SEM)**. The alternative CB-SEM method has been labeled as hard modeling because of its more stringent assumptions in terms of data distribution and requiring much larger sample sizes (e.g., Falk & Miller, 1992). Importantly, for PLS-SEM "it is not the concepts nor the models nor the estimation techniques which are 'soft,' only the distributional assumptions" (Lohmöller, 1989, p. 64). While both methods were developed at about the same time, CB-SEM became much more widely applied because of its early availability through the LISREL software in the late 1970s. In contrast, the first software for PLS-SEM was LVPLS, which appeared in the mid-1980s (Lohmöller, 1984, 1987). But this initial software was not very user-friendly, and it was not until Chin's (1994) PLS-Graph software in the mid-1990s that PLS-SEM experienced wider application. With the release of SmartPLS 2 in 2005 (Ringle, Wende, & Will, 2005), SmartPLS 3 in 2015 (Ringle, Wende, & Becker, 2015), and SmartPLS 4 in 2022 (Ringle, Wende, & Becker, 2022), PLS-SEM applications increased exponentially (Sarstedt & Cheah, 2019), as evidenced by the popularity of the terms "PLS-SEM" and "PLS path modeling" in the Web of Science database in terms of articles published and citations (Exhibit 1.1).

Over the last two decades, there have been numerous introductory articles on the method (e.g., Chin, 1998; Haenlein & Kaplan, 2004; Hair, Risher, Sarstedt, & Ringle, 2019; Nitzl & Chin, 2017; Rigdon, 2013; Roldán & Sánchez-Franco, 2012;

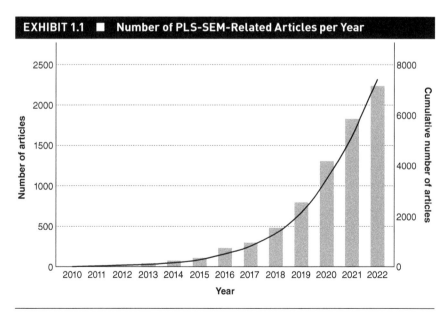

EXHIBIT 1.1 ■ Number of PLS-SEM-Related Articles per Year

Note: Number of articles returned from the Web of Science database for the search terms "PLS-SEM" and "PLS path modeling" beginning from 2010.

Tenenhaus, Esposito Vinzi, Chatelin, & Lauro, 2005) as well as review articles examining how researchers in business and related fields have applied it (Exhibit 1.2). The usage of PLS-SEM also expanded into other research areas, such as agriculture, engineering, environmental sciences and ecology, geography, and psychology (Sarstedt, 2019).

With the increasing maturation of the PLS-SEM field (Hwang, Sarstedt, Cheah, & Ringle, 2020; Khan et al., 2019), researchers can draw on a much greater repertoire

EXHIBIT 1.2 ■ Review Articles on PLS-SEM Usage	
Discipline	*References*
Accounting	Lee, Petter, Fayard, & Robinson (2011)
	Nitzl (2016)
Entrepreneurship	Manley, Hair, Williams, & McDowell (2020)
Family business	Sarstedt, Ringle, Smith, Reams, & Hair (2014)
Higher education	Ghasemy, Teeroovengadum, Becker, & Ringle (2020)
Hospitality and tourism	Ali, Rasoolimanesh, Sarstedt, Ringle, & Ryu (2018)
	Do Valle & Assaker (2016)
	Usakli & Kucukergin (2018)
Human resource management	Ringle, Sarstedt, Mitchell, & Gudergan (2020)
International business research	Richter, Sinkovics, Ringle, & Schlägel (2016)
Knowledge management	Cepeda-Carrion, Cegarra-Navarro, & Cillo (2019)
Management	Hair, Sarstedt, Pieper, & Ringle (2012)
Marketing	Hair, Sarstedt, Ringle, & Mena (2012)
	Guenther, Ringle, Zaefarian, & Cartwright (2023)
	Sarstedt, Hair, Pick, Liengaard, Radomir, & Ringle (2022)
Management information systems	Hair, Hollingsworth, Randolph, & Chong (2017)
	Ringle, Sarstedt, & Straub (2012)
Operations management	Bayonne, Marin-Garcia, & Alfalla-Luque (2020)
	Peng & Lai (2012)
Psychology	Willaby, Costa, Burns, MacCann, & Roberts (2015)
Quality management	Magno, Cassia, & Ringle (2022)
Software engineering	Russo & Stol (2021)
Supply chain management	Kaufmann & Gaeckler (2015)

of advanced modeling, analysis techniques, and robustness checks (e.g., Sarstedt, Ringle, Cheah, Ting, Moisescu, & Radomir, 2020) to support their conclusions' validity and to identify more complex relationship patterns. For example, methodological research has made considerable progress in the treatment of observed heterogeneity in the context of moderator analysis (Becker et al., 2023; Memon et al., 2019), multigroup analysis (Chin & Dibbern, 2010; Klesel, Schuberth, Niehaves, & Henseler, 2021), and invariance assessment (Henseler, Ringle, & Sarstedt, 2016). Researchers have also developed novel latent class procedures (Becker, Rai, Ringle, & Völckner, 2013; Schlittgen, Ringle, Sarstedt, & Becker, 2016) and guidelines for their use (Sarstedt, Ringle, & Hair, 2021). Furthermore, progress has been made in the specification, estimation, and validation of higher order models (Becker, Cheah, Gholamzade, Ringle, & Sarstedt, 2023; Sarstedt, Hair, Cheah, Becker, & Ringle, 2019) and nonlinear effects (Basco, Hair, Ringle, & Sarstedt, 2021). With these extensions, PLS-SEM has become a full-fledged estimator for latent variable models and is capable of handling many modeling problems social sciences researchers face today.

MODEL SPECIFICATION

Model specification in PLS-SEM involves two types of sub-models: the structural model and the measurement models. The **structural model** (also referred to as the **inner model**) specifies the relationships between the constructs. Constructs that act only as independent variables are referred to as **exogenous constructs**, whereas those that act as dependent variables are called **endogenous constructs**. The relationships between constructs are typically visualized in a path model, such as the one shown in Exhibit 1.3.

In this path model, Y_3 and Y_4 act as endogenous constructs, while Y_1 and Y_2 are exogenous. The endogenous constructs have **error terms** z_3 and z_4 connected to them,

EXHIBIT 1.3   ■   Partial Least Squares Path Model Example

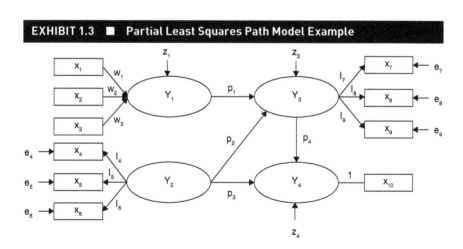

which represent the unexplained variance (i.e., the difference between the model's in-sample prediction and actual construct scores). The exogenous construct Y_1 also has an error term (z_1) but in PLS-SEM, this error term is zero by default because of the way the method treats the measurement model of this particular construct, which is formative in nature (Diamantopoulos, 2011). Therefore, the error term is typically omitted in the display of a PLS path model. The other exogenous construct Y_2 is based on a reflective measurement model (i.e., the arrows point from the construct to its indicators) and therefore has no error term attached to it.

The structural model relationships in Exhibit 1.3 can be expressed using the following formulas:

$$Y_3 = p_1 \cdot Y_1 + p_2 \cdot Y_2 + z_3 \text{ and}$$
$$Y_4 = p_3 \cdot Y_2 + p_4 \cdot Y_3 + z_4$$

The **measurement models** (also referred to as **outer models**) express the relationships between each construct and its indicators. There are two broad ways to conceptualize measurement models from a measurement theory perspective. The first approach is referred to as reflective measurement. In a **reflective measurement model**, the indicators are considered to be error-prone manifestations of an underlying construct. That is, the relationships between the construct and the indicators are likely to include errors. Reflective measurement model relationships are represented as arrows going from the construct to its indicators. For example, the construct Y_2 in Exhibit 1.3 has a reflective measurement model, which can be expressed in the following way:

$$x_4 = l_4 \cdot Y_2 + e_4,$$
$$x_5 = l_5 \cdot Y_2 + e_5, \text{ and}$$
$$x_6 = l_6 \cdot Y_2 + e_6.$$

The terms l_4 to l_6 are the standardized indicator loadings, which are calculated from three bivariate regressions in which each reflective indicator x_4 to x_6 acts as a dependent variable and the construct Y_2 as independent variable (i.e., l_4 to l_6 simply represent the bivariate correlations between construct Y_2 and each of its indicators x_4 to x_6); e_4 to e_6 are the error terms representing the unexplained variance in each regression model. Note that there is no regression intercept, as PLS-SEM works with standardized data (i.e., the intercept is zero). **Reflective indicators** (sometimes referred to as **effect indicators** in the psychometric literature) can be viewed as a representative sample of all the possible items available in the conceptual domain of the construct (Nunnally & Bernstein, 1994). Since a reflective measurement model requires that all items reflect the same construct, indicators associated with a particular construct should be highly correlated with each other. In addition, individual items should be interchangeable, and any single item can generally be left out without changing the meaning of the construct, as long as the construct has sufficient reliability. The fact that the relationship goes from the construct to its indicators implies that if the evaluation of the latent trait changes (e.g., because of a change in the standard of comparison), all indicators will change simultaneously—at least to some extent.

The other type of measurement model is formative measurement. In a **formative measurement model**, the indicators form the construct using linear combinations. A change in an indicator's value due to, for example, a change in a respondent's assessment of the trait being captured by the indicator changes the value of the construct. That is, variation in the indicators precedes variation in the constuct (Borsboom, Mellenbergh, & van Heerden, 2003), which means that, by definition, constructs with a formative measurement model are inextricably tied to their measures (Diamantopoulos, 2006). Besides the difference in the relationship between indicator(s) and construct, formative measurement models do not require correlated indicators. Formative measurement model relationships are represented by arrows leading from the indicators to the construct.

Researchers distinguish between two types of indicators in the context of formative measurement: composite and causal indicators. **Composite indicators** largely correspond to the above definition of formative measurement models in that they are combined in a linear way to form a variate, which is also referred to as a composite in the context of SEM (Bollen, 2011; Bollen & Bauldry, 2011). More precisely, the indicators fully form the construct (i.e., the construct's R^2 value is 1.0), which means the construct has zero error. Composite indicators have often been used to measure **artifacts**, which can be understood as human-made concepts (Henseler, 2017). Examples of such artifacts in marketing include the retail price index or the marketing mix (Hair, Sarstedt, & Ringle, 2019). However, composite indicators can also be used to measure attitudes, perceptions, and behavioral intentions (Sarstedt, Hair, Ringle, Thiele, & Gudergan, 2016; Rossiter, 2011; Rossiter, 2016), provided the indicators have conceptual unity in accordance with a clear theoretical definition (Gilliam & Voss, 2013). The PLS-SEM algorithm relies solely on the concept of composite indicators because of the way the algorithm estimates formative measurement models (e.g., Diamantopoulos, 2011).

Causal indicators also form the construct, but this type of measurement acknowledges that it is unlikely that any set of causal indicators can fully capture every aspect of a latent phenomenon (Bollen & Diamantopoulos, 2017; Diamantopoulos & Winklhofer, 2001). Therefore, constructs measured with causal indicators have an error term, which is assumed to capture all the other causes of the latent variable not included in the model (Diamantopoulos, 2006). The use of causal indicators is prevalent in CB-SEM, which—at least in principle—allows for explicitly defining the error term of a formatively measured latent variable. However, the nature of this error term is ambiguous, as its magnitude partly depends on other constructs embedded in the model and their measurement quality (Sarstedt, Hair, Ringle, Thiele, & Gudergan, 2016).

The path model in Exhibit 1.3 has one formatively measured construct, Y_1, which PLS-SEM estimates using composite indicators. The corresponding measurement model can therefore be expressed as follows:

$$Y_1 = w_1 \cdot x_1 + w_2 \cdot x_2 + w_3 \cdot x_3 + z_1,$$

in which $z_1 = 0$.

Rather than using multiple items to measure a construct, researchers sometimes opt for a **single-item measurement**. PLS-SEM proves valuable in this respect, as the method does not encounter identification problems when using less than three items in a measurement model as it is the case with CB-SEM. Single items have practical advantages, such as ease of application, brevity, and lower costs associated with their use. Unlike long and complicated scales, which sometimes result in a lack of understanding and mental fatigue for respondents, single items promote higher response rates since the questions can be easily and quickly answered (Fuchs & Diamantopoulos, 2009; Sarstedt & Wilczynski, 2009). However, single-item measures do not offer more for less. For instance, when partitioning the data into groups, researchers have fewer options since scores from only a single variable are available to partition the data. Similarly, information is available from only a single measure instead of several measures when using imputation methods to deal with missing values.

More importantly, from a psychometric perspective, single-item measures do not allow for the removal of measurement error (as is the case with multiple items), which generally decreases the measure's reliability. Note that, contrary to commonly held beliefs, single-item reliability can be estimated (e.g., Cheah, Sarstedt, Ringle, Ramayah, & Ting, 2018; Loo, 2002; Wanous, Reichers, & Hudy, 1997). In addition, opting for single-item measures in most empirical settings is a risky decision when it comes to predictive validity considerations. Specifically, the set of circumstances that would favor the use of single-item over multi-item measures is very unlikely to be encountered in practice. Finally, social sciences scholars often include complex constructs in their theoretical models, such as trust, commitment, cooperation, and satisfaction. Using a single-item measure to represent such complex attitudinal or behavioral concepts can reduce the validity of the construct. According to guidelines by Diamantopoulos, Sarstedt, Fuchs, Kaiser, and Wilczynski (2012), single-item measures should be considered only in situations when (1) small sample sizes are present (i.e., $N < 50$), (2) path coefficients (i.e., the coefficients linking constructs in the structural model) of 0.30 and lower are expected, (3) items of the originating multi-item scale are highly homogeneous (i.e., inter-item correlations > 0.80, Cronbach's alpha > 0.90), and (4) the items are semantically redundant. Against this background, we generally advise against the use of single items for construct measurement. For further discussions on the efficacy of single-item measures, see, for example, Kamakura (2015) and Sarstedt, Diamantopoulos, Salzberger, and Baumgartner (2016).

MODEL ESTIMATION

The Original PLS-SEM Algorithm

With an adequate sample (see Hair, Hult, Ringle, & Sarstedt, 2022, Chapter 1, for further details), which meets the minum sample size, researchers can use the PLS-SEM method for the model estimation (see also Ringle, Sarstedt, Sinkovics, & Sinkovics, 2023). Model estimation in PLS-SEM draws on a three-stage approach that belongs to the family of (alternating) least squares algorithms (Mateos-Aparicio, 2011). Exhibit 1.4 illustrates the **PLS-SEM algorithm** as presented by Lohmöller (1989, Chapter 2). Henseler et al. (2012) offer a graphical illustration of its stages. The algorithm starts with an initialization stage in which it establishes preliminary construct scores. To compute these scores, the algorithm typically uses unit weights (i.e., 1) for all indicators in the measurement models (Hair et al., 2022).

Stage 1 of the PLS-SEM algorithm iteratively determines the inner weights (i.e., the path coefficients) and construct scores employing a four-step procedure. Step #1 uses the initial construct scores from the initialization of the algorithm to determine

EXHIBIT 1.4 ■ The Basic PLS-SEM Algorithm (adapted from Lohmöller 1989, p. 29)	
Initialization	
Stage 1:	**Iterative estimation of weights and construct scores**
	Starting at Step #4, repeat Steps #1 to #4 until convergence is obtained.
#1	Inner weights (here obtained by using the factor weighting scheme)
	$b_{ji} = \begin{cases} cov(Y_j; Y_i) & \text{if } Y_j \text{ and } Y_i \text{ are adjacent} \\ 0 & 0 \text{ else} \end{cases}$
#2	Inside approximation
	$\bar{Y}_j := \sum_i b_{ji} Y_i$
#3	Outer weights; solve for
	$\bar{Y}_{jn} = \sum_k \bar{w}_{k_j} x_{kn} + d_{jn}$ in a Mode A block
	$x_{kn} = \bar{w}_{k_j} \bar{Y}_{jn} + e_{kn}$ in a Mode B block
#4	Outside approximation
	$Y_{jn} := \sum_k \bar{w}_{k_j} x_{kn}$
Stage 2:	**Estimation of outer weights, outer loadings, and path coefficients**
Stage 3:	**Estimation of location parameters**

the inner weights b_{ji} between the adjacent constructs Y_j (i.e., the dependent one) and Y_i (i.e., the independent one) in the structural model. Please note that we use an extended nomenclature compared to our discussion of model specification. For example, the inner weights b_{ji}, which are used as provisional estimates of the path coefficients, have two indices, i and j, representing the independent construct and dependent construct of the corresponding relationship. The literature suggests different approaches to determining the inner weights (Chin, 1998; Lohmöller, 1989; Tenenhaus, Esposito Vinzi, Chatelin, & Lauro, 2005). In the **factor weighting scheme**, the inner weight corresponds to the covariance between Y_j and Y_i and is set to 0 if the constructs are unconnected. The **path weighting scheme** considers the direction of the inner model relationships (Lohmöller, 1989, Chapter 2). Chin (1998, p. 309) notes that the path weighting scheme "attempts to produce a component that can both ideally be predicted (as a predictand) and at the same time be a good predictor for subsequent dependent variables." As a result, the path weighting scheme leads to slightly higher R^2 values in the endogenous constructs compared to the other schemes and should therefore be preferred. In most instances, however, the choice of the inner weighting scheme has very little bearing on the results (Lohmöller, 1989, Chapter 2; Noonan & Wold, 1982).

Step #2, the inside approximation, computes proxies for all constructs $\tilde{Y}_j$ by using the weighted sum of its adjacent constructs' scores Y_i. Then, for all the indicators in the measurement models, Step #3 computes new outer weights, which indicate the strength of the relationship between each construct $\tilde{Y}_j$ and its corresponding indicators. To do so, the PLS-SEM algorithm uses two different estimation modes (Exhibit 1.4). When using **Mode A** (i.e., **correlation weights**), the bivariate correlation between each indicator and the construct determines the outer weights. In contrast, **Mode B** (i.e., **regression weights**) computes indicator weights by regressing each construct on its associated indicators.

By default, estimation of reflectively specified constructs draws on Mode A, whereas PLS-SEM uses Mode B for formatively specified constructs. However, Cho et al. (2023) show that this reflex-like use of Mode A and Mode B is not optimal when using PLS-SEM for prediction purposes. Their simulation study indicates that Mode A provides higher degrees of out-of-sample predictive power in situations commonly encountered in empirical research (see also Becker, Rai, & Rigdon, 2013). Exhibit 1.4 shows the formal representation of these two modes, where x_{kjn} represents the raw data for indicator k ($k = 1, \ldots, K$) of construct j ($j = 1, \ldots, J$) and observation n ($n = 1, \ldots, N$); $\tilde{Y}_{jn}$ are the construct scores from the inside approximation in Step #2, $\tilde{w}_{kj}$ are the outer weights from Step #3, d_{jn} is the error term from a bivariate regression, and e_{kjn} is the error term from a multiple regression. The updated weights from Step #3 (i.e., $\tilde{w}_{kj}$) and the indicators (i.e., x_{kjn}) are linearly combined to update the constructs scores (i.e., Y_{jn}) in Step #4 (outside approximation). Note that the PLS-SEM algorithm uses standardized data as input and always standardizes the generated construct scores in Step #2 and Step #4. After Step #4, a new iteration starts. The algorithm terminates when the weights obtained from Step #3 change marginally from one iteration to the next (typically $1 \cdot 10^{-7}$), or when the maximum number of iterations is achieved (typically 300).

Stages 2 and 3 use the final construct scores from Stage 1 as input for a series of ordinary least squares regressions. These regressions compute the final outer loadings, outer weights, and path coefficients as well as related elements such as indirect, and total effects, R^2 values of the endogenous constructs, and the indicator and construct correlations (Lohmöller, 1989, Chapter 2).

The Weighted PLS-SEM Algorithm

An extension of the PLS-SEM approach is the **weighted PLS-SEM (WPLS)** algorithm (Becker & Ismail, 2016). This modified version of the original PLS-SEM algorithm enables researchers to match sample and population structure (Cheah, Roldán, Ciavolino, Ting, and Ramayah, 2021).

When estimating a PLS path model, researchers typically seek to draw inferences about the population of interest. An important requirement for such inferences is that the sample is representative of the population. Probability sampling methods, such as simple random sampling or cluster sampling, meet this requirement as every member of a population has an equal probability of being selected in the sample (Sarstedt & Mooi, 2019, Chapter 4). In the probability sampling case, every observation in the sample would have the same weight in the PLS-SEM analysis. In practice, however, population members are often not equally likely to be included in the sample, for example, because of the use of non-probability sampling methods, such as quota sampling, which is the norm for social sciences research. To adjust for resulting biases, researchers may use **sampling weights** (also referred to as post-stratification weights) that assign the observations different relevance in the parameter estimation process (Sarstedt, Bengart, Shaltoni, & Lehmann, 2018). For example, if a population consists of an equal share of males and females but the sample comprises 60% males and 40% females, sampling weights ensure that females are weighted more strongly than males in the parameter estimation.

Different from the original PLS-SEM algorithm, the WPLS algorithm considers sampling weights v_n in the calculation of the mean, the variance, and covariance (correlation) of the construct scores in each iteration. For example, indicator standardization should rely on the weighted mean $\hat{\bar{x}}$ and weighted variance $\widehat{var(x)}$, defined as follows:

$$\hat{\bar{x}} = \frac{\sum_{n=1}^{N} v_n x_n}{\sum_{n=1}^{N} v_n}$$

$$\widehat{var(x)} = \frac{\sum_{n=1}^{N} v_n \left(x_n - \hat{\bar{x}}\right)^2}{\left(\sum_{n=0}^{N} v_n\right)},$$

whereby the hat symbolizes that these are weighted results. Similarly, the correlation weights used in Mode A draw on the weighted covariances, while the regression weights in Mode B and the inner model weights should use the weighted standardized data as input. For example, the inner model weights are given by:

$$b_{ji} = cor_v(Y_i)^{-1} cor_v\left(Y_i, \tilde{Y}_j\right).$$

The effect of these corrections is that all the calculations during the iterative PLS-SEM algorithm (e.g., path coefficient estimates) are weighted with the sampling weights while retaining all information from the original data set in the model. As a result, WPLS provides more accurate average population model parameter estimates than the basic PLS-SEM algorithm when appropriate sampling weights are available. Becker and Ismail (2016) and Cheah, Roldán, Ciavolino, Ting, and Ramayah (2021) provide more details on WPLS.

Consistent PLS-SEM

The **consistent PLS-SEM (PLSc-SEM)** algorithm performs a correction of reflective constructs' correlations to make the results consistent with a common factor model (Dijkstra 2010, 2014; Dijkstra & Henseler 2015; Dijkstra & Schermelleh-Engel 2014). In principle, the correction builds on Nunnally's (1978) well-known correction for attenuation formula. As such, PLSc-SEM follows a composite modeling logic and modifies the PLS-SEM results to mimic a common factor model (Sarstedt, Hair, Ringle, Thiele, & Gudergan, 2016) and, thus, the results generated by CB-SEM (Jöreskog, 1978; Jöreskog & Wold, 1982).

Specifically, PLSc-SEM's objective is to compute the disattenuated (i.e., consistent) correlation $r^c_{Y_1,Y_2}$ between two constructs Y_1 and Y_2, which is the modified version of their original correlation (r_{Y_1,Y_2}), corrected for measurement error. To do so, PLSc-SEM divides the original construct correlation r_{Y_1,Y_2} by the geometric mean of the constructs' reliabilities, measured using the **reliability coefficient ρ_A**.

PLSc-SEM follows a four-step approach (Exhibit 1.5). In Step 1, the traditional PLS-SEM algorithm is run. These results are then used in Step 2 to calculate the reliability ρ_A of all reflectively measured constructs in the PLS path model (Dijkstra, 2014; Dijkstra & Henseler, 2015b). For formatively measured constructs and single-item constructs, ρ_A is set to 1. In Step 3, the consistent reliabilities of all constructs from Step 2 are used to correct the inconsistent correlation matrix of the constructs

EXHIBIT 1.5 ■ The Four-Step PLSc-SEM Procedure

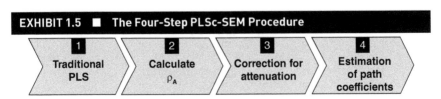

1	2	3	4
Traditional PLS	Calculate ρ_A	Correction for attenuation	Estimation of path coefficients

obtained in Step 1. More precisely, researchers obtain the consistent correlation between two constructs by dividing their correlation from Step 1 by the geometric mean (i.e., the square root of the product) of their reliabilities ρ_A. This correction applies to all correlations of reflectively measured constructs. The correlation of two formative and/or single-item constructs remains unchanged. The correction for attenuation only applies when at least one reflectively measured construct with a consistent reliability ρ_A smaller than 1 is involved in the correlation between two constructs in the PLS path model. In Step 4, the consistent correlation matrix of the constructs allows re-estimating all model relationships yielding consistent path coefficients, corresponding R^2 values, and outer loadings. Note that significance testing in PLSc-SEM requires running an adjusted bootstrapping routine, which has also been implemented in extant PLS-SEM software.

In practical applications, PLSc-SEM results can be substantially influenced by low reliability levels of the constructs. As a result, the standardized path coefficients produced by PLSc-SEM can become very high (in some situations considerably larger than 1). Moreover, in more complex PLS path models, collinearity among the constructs has a strong negative impact on the PLSc-SEM results. In some instances, the structural model relationships become very small. Finally, bootstrapping of PLSc-SEM results frequently produces inadmissible solutions or can yield extreme outcomes, which result in high standard errors in certain relationships, increasing the Type II error rate.

In light of these limitations, the question arises as to when researchers should use PLSc-SEM. The PLSc-SEM approach is appropriate when researchers assume the data are obtained from a common factor model (Bollen, 2011; Bollen & Bauldry, 2011); we will discuss this model type in detail in the following section. In that case, the objective is to mimic CB-SEM results by assuming the construct can be adequately represented by the common variance of its indicators. Simulation studies for such models reveal that CB-SEM and PLSc-SEM return almost identical results of the estimated coefficients (Dijkstra & Henseler, 2015b). While CB-SEM and PLSc-SEM have approximately the same accuracy of estimated parameters and statistical power, PLSc-SEM retains most of PLS-SEM's advantageous features. Among others, PLSc-SEM does not rely on distributional assumptions, can handle complex models, is less affected by incorrect specifications in subparts of the model, and will not encounter convergence problems. At the same time, however, the correction for attenuation in Step 3 changes the path coefficient estimates, which have been derived from the original PLS-SEM estimation and maximize the endogenous constructs' explained variance. As a consequence, any assessment of the model's predictive power using the modified path coefficient estimates is inconsistent with the original PLS-SEM estimation. Considering that research has emphasized the causal-predictive nature as an integral part of PLS-SEM and a key distinguishing feature from CB-SEM, this limitation is highly problematic. In addition, Sarstedt, Hair, Ringle, Thiele, and Gudergan (2016) show that the bias produced by PLSc-SEM is considerably higher than the one produced by CB-SEM when erroneously using the method on data that stem from a composite model population.

In light of these limitations, Hair, Sarstedt, and Ringle (2019, p. 567) conclude that PLSc-SEM "adds very little to existing knowledge of SEM" and that researchers should revert to the widely recognized and accepted CB-SEM approach when estimating common factor models. Nevertheless, PLSc-SEM is an alternative to the standard CB-SEM estimation when attempting to estimate under-identified models or when convergence problems occur. The same limitations apply to Bentler & Huang's (2014) PLSe2 method, which builds on the PLSc-SEM results but applies a generalized least squares covariance structure estimation on the modified correlation matrix, and Yuan, Wen, & Tang's (2020) Cronbach's α based approach. As such, the method does not unite the advantages of PLS-SEM and CB-SEM as suggested by some researchers (Ghasemy, Jamil, & Gaskin, 2021).

PRINCIPLES OF PLS-SEM

Several considerations are important when applying PLS-SEM, which have their roots in the method's characteristics. While PLS-SEM is a distinct statistical method (Schuberth, Zaza, & Henseler, 2023), some of its characteristics need to be explained by contrasting it to the CB-SEM method, which has long dominated social sciences research (Babin, Hair, & Boles, 2008). We first discuss the measurement philosophy underlying PLS-SEM, followed by aspects related to the way the method estimates model parameters and ensuing biases. We then summarize the implications regarding the recommended situations for the application of each method based on strengths and limitations.

Philosophy of Measurement

A crucial conceptual difference between PLS-SEM and CB-SEM relates to the way each method calculates the scores of all constructs included in a model. CB-SEM represents a **common factor-based SEM** method that considers the constructs as common factors that explain the covariances between all indicators included in the theoretical model. The underlying assumption is that the covariances (or **common variance**) of a set of indicators can in principle, be perfectly explained by the existence of one unobserved variable (the common factor) and an individual random error (Spearman, 1927; Thurstone, 1947). The **common factor model** approach is consistent with the measurement philosophy underlying reflective measurement in which the indicators and their covariations are regarded as manifestations of the underlying construct. To estimate the model parameters, CB-SEM minimizes the divergence between the empirical (observed) covariance matrix and the covariance matrix implied (estimated) by the model given a certain set of parameter estimates (Hair, Black, Babin, & Anderson, 2019).

In contrast, PLS-SEM runs partial regressions to obtain construct scores that minimize the residuals (error variances) in the relationships between composites and indicators (i.e., in the measurement models) as well as those between composites (i.e., in the structural model)—see Tenenhaus, Esposito Vinzi, Chatelin, and Lauro (2005). In doing so, PLS-SEM linearly combines the indicators of each construct's measurement model to form composite variables. PLS is therefore considered a **composite-based SEM** method (Hwang, Sarstedt, Cheah, & Ringle, 2020). In estimating the composite scores, the PLS-SEM algorithm weights each indicator individually. The indicator weights reflect each indicator's importance in forming the composite. That is, indicators with a larger weight contribute more strongly to forming it. In addition, unlike CB-SEM, in which model solutions are based only on common variance, PLS-SEM solutions are derived from total variance, which consists of both common and unique variance.

Moreover, PLS-SEM composite scores are superior to **sum scores**, which unit-weight each item with a coefficient of 1 (or any other arbitrary value so long as it is constant) for representing constructs (McNeish & Wolf, 2020). For example, with reflective measurement models, the PLS-SEM weights are also indicative of each indicator's degree of measurement error. Indicators with high degrees of measurement error have smaller weights, thereby contributing less to forming the composite variable (Hair, Hult, Ringle, & Sarstedt, 2021). The process of applying weighted composites of indicator variables makes PLS-SEM superior to multiple regression and other statistical models using sum scores. If multiple regression with sum scores is used, the researcher assumes an equal weighting of all construct indicators, which means each indicator contributes equally to forming the composite, thereby not considering measurement error in calculating the construct scores (Hair & Sarstedt, 2019; Henseler et al., 2014). As a result, the use of sum scores produces higher parameter bias when the indicator weights differ, as is typically the case in applied research. Moreover, sum scores SEM approaches produce lower statistical power compared to differential indicator weights produced by PLS-SEM (Hair, Hult, Ringle, Sarstedt, & Thiele, 2017). McNeish and Wolf (2020) summarize the empirical and conceptual shortcomings of sum scores estimation, noting equal weights (1) generate unrealistic expectations about the population (data-generating) model by enforcing unnatural constraints on the empirical model; (2) hinder rigorous and accurate psychometric assessments by ignoring measurement theory in its entirety; (3) adversely affect construct validity and reliability; and (4) often result in vastly different conclusions due to inaccurate coefficient estimation. In addition, because virtually all the psychometric scales used in business research have been validated under the assumption of differentiated weights, using equal weights when applying these scales is categorically inappropriate because doing so imposes a different model from the initial validated one.

Parameter Estimation Accuracy

The philosophy of measurement assumed by PLS-SEM has important implications for any statement regarding its parameter accuracy (i.e., the degree to which the method produces accurate results). At the beginning of the method's development and use, researchers noted that PLS estimation is "deliberately approximate" (Hui & Wold 1982, p. 127) to common factor-based SEM, a characteristic that has come to be incorrectly known as the **PLS-SEM bias** (e.g., Marcoulides, Chin, & Saudners, 2012). Several studies have used simulations to demonstrate the alleged PLS-SEM bias (e.g., Goodhue, Lewis, & Thompson, 2012; McDonald, 1996; Rönkkö & Evermann, 2013), which supposedly manifests itself in measurement model estimates that are higher and structural model estimates that are lower compared to the prespecified values. The studies conclude that parameter estimates will approach what has been labeled the "true" parameter values when both the number of indicators per construct and sample size increase (Hui & Wold, 1982). All these simulation studies used CB-SEM as the benchmark against which the PLS-SEM estimates were evaluated based on the assumption that the results should be the same. However, such assessments can be expected to include bias since PLS-SEM is a composite-based approach that uses the total variance to estimate parameters (e.g., Schlittgen, Sarstedt, & Ringle, 2020; Schneeweiß, 1991; Tenenhaus, Esposito Vinzi, Chatelin, & Lauro, 2005). Not surprisingly, the very same issues apply when composite models are used to estimate CB-SEM results. In fact, Sarstedt, Hair, Ringle, Thiele, and Gudergan (2016) show that the biases produced by the CB-SEM methods are far more severe than those of PLS-SEM, when applying the method to the wrong type of model (i.e., estimating composite models with CB-SEM vs. estimating common factor models with PLS-SEM). Cho, Sarstedt, and Hwang (2022) recently confirmed these findings in a more complex simulation design and conceptually compared common factor and composite models, clarifying their similarities and differences. When acknowledging the different nature of the construct measures, most of the arguments voiced by critics of the PLS-SEM method (Rönkkö, McIntosh, Antonakis, & Edwards, 2016) are no longer an issue (Cook & Forzani, 2020; Cook & Forzani, 2023). Yuan and Fang (2022) raise additional conceptual concerns regarding the assumed PLS-SEM bias. They note that the parameter values in a CB-SEM analysis depend on the researcher's way of fixing the scale (e.g., fixing an indicator loading to unity)—a necessary step in any CB-SEM analysis. As a consequence, "the population values of the model parameters under [CB-]SEM are artificial," which implies that any resulting bias "does not enjoy a clear substantive interpretation." Cook and Foranzi (2023) recently echoed this notion.

Apart from these conceptual concerns, simulation studies show the differences between PLS-SEM and CB-SEM estimates when assuming the latter as a standard of comparison are very small, provided the measurement models meet minimum recommended standards in terms of measurement quality (i.e., reliability and validity). Specifically, when the measurement models have four or more indicators and the indicator loadings meet the common standards (≥ 0.708), there are practically no differences between the two methods in terms of parameter accuracy (e.g., Reinartz, Haenlein, & Henseler, 2009; Sarstedt, Hair, Ringle, Thiele, & Gudergan, 2016). Thus, the extensively discussed and supposed PLS-SEM bias is of no practical relevance for almost all SEM applications (e.g., Binz-Astrachan, Patel, & Wanzenried, 2014; Sharma, Liengaard, Sarstedt, Hair, & Ringle, 2023).

A more fundamental question is, however, whether it is reasonable to view common factor models as the universal measurement benchmark. Research casts considerable doubt on this premise, as common factors derived in CB-SEM are also not necessarily equivalent to the theoretical concepts that are the focus of research (Rigdon, 2012; Rigdon, Sarstedt, & Ringle, 2017; Rossiter, 2011; Sarstedt, Hair, Ringle, Thiele, & Gudergan, 2016). The reason for concern regarding the use of CB-SEM measurement and structural model results as a benchmark is that measurement in CB-SEM defines validity purely on the grounds of the relationships between the constructs and their indicators as measured by common variance only. For example, strong loadings in a measurement model suggest the construct converges into its indicators, offering evidence for convergent validity. However, constructs merely serve as representations or proxies of conceptual variables in a statistical model (Rigdon & Sarstedt, 2022; Sarstedt, Hair, Ringle, Thiele, & Gudergan, 2016). The actual entities of interest are the conceptual variables, which represent broad ideas or thoughts about abstract concepts that researchers establish and propose to measure in their research (e.g., Bollen, 2002). The relationships between the constructs and the conceptual variables, however, remain unknown since measurement error, as defined in SEM, only relates to the relationships between indicators and common factors or composites (Hair & Sarstedt, 2019). Exhibit 1.6 illustrates the relationships between conceptual variables, constructs, and indicators.

Even when considering perfect model fit, it is unreasonable to assume the constructs are equivalent to the conceptual variables they seek to represent (Cliff, 1983; MacCallum & Browne, 2007; Michell, 2013). The reason is any measurement includes **metrological uncertainty**, which refers to the dispersion of the measurement values attributed to the object or concept being measured (JCGM/WG1, 2008). Numerous sources contribute to metrological uncertainty, such as definitional uncertainty or limitations related to the measurement scale design, which go well beyond the simple standard errors produced by CB-SEM analyses (Hair & Sarstedt, 2019; Rigdon, Sarstedt, & Becker, 2020). One contributing factor to uncertainty is **factor (score) indeterminacy**, which means an infinite number of different sets of construct scores that will fit the model equally well are possible (Guttman, 1955; Schönemann & Wang, 1972). Factor indeterminacy produces a band of uncertainty in the relationship (i.e., the correlation) between a common factor inside the model and any variable outside the model (Steiger, 1979). Since the conceptual variable itself is not part of the model (i.e., it is outside the model), this range of correlation values applies as well to the relationship between an indeterminant common factor and the conceptual variable it is designed to represent (Exhibit 1.6). For example, the presence of factor indeterminacy implies the correlation between the common factor in the statistical model labeled, for example, customer satisfaction and the actual customer satisfaction is uncertain—even when the model shows a perfect fit (Rigdon & Sarstedt, 2022). To better grasp this concept, consider a model with four moderately correlated (0.3) common factors, each measured with three indicators and all with 0.7 loadings. In such a setting, the correlation between a common factor (e.g., the construct labeled *customer satisfaction*) and the corresponding

EXHIBIT 1.6 ■ Error Framework for Conceptual Variables, Proxies, and Indicators

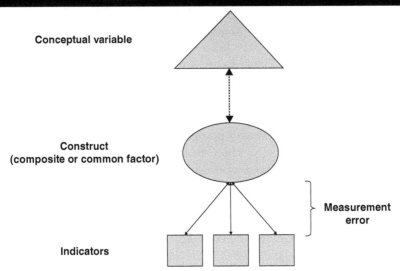

conceptual variable (e.g., the actual *customer satisfaction*) lies in a range that has a width of 0.495 (Rigdon, Becker, & Sarstedt 2019). This means the range of uncertainty covers 0.495/2 = 24.75% of the total possible range of correlation (see Table 2 in Rigdon, Becker, & Sarstedt, 2019). Researchers' common practice of restricting the number of indicators per construct to improve model fit in CB-SEM analyses further increases this uncertainty (Hair, Matthews, Matthews, & Sarstedt, 2017; Rigdon, Becker, & Sarstedt, 2019).

While these issues do not necessarily imply that PLS-SEM is superior (Rigdon & Sarstedt, 2022), they cast considerable doubt on the assumption of some researchers that CB-SEM constitutes the gold standard when measuring unobservable concepts (e.g., Rhemtulla, van Bork, & Borsboom, 2020; Rigdon, 2016). Instead, researchers should recognize that all constructs are merely approximations of or proxies for conceptual variables (Exhibit 1.6), independent from how they were estimated (e.g., Rigdon, Sarstedt, & Ringle, 2017; Wickens, 1972). That is, constructs should be viewed as "something created from the empirical data which is intended to enable empirical testing of propositions regarding the concept" (Rigdon 2012, pp. 343–344). Hence, irrespective of the quality with which a conceptual variable is theoretically substantiated and operationally defined, and the rigor that encompasses measurement model development, any measurement in structural equation models produces only proxies for conceptual variables (Rigdon, 2012). This assessment is in line with the proliferation of all sorts of instruments that claim to measure essentially the same concept, although often with little chance to convert one instrument's measures into any other instrument's measures (Salzberger, Sarstedt, & Diamantopoulos, 2016). For example, research and practice have proposed a multitude of measurement instruments for corporate reputation, which rest on the same definition of the concept but differ fundamentally in terms of their underlying conceptualizations and measurement items (Sarstedt, Wilczynski, & Melewar, 2013). Similarly, Bergkvist and Langner (2017, 2019) find considerable heterogeneity in the operationalizations of common advertising constructs, such as attitude toward the ad, attitude toward the brand, ad credibility, ad irritation, and brand purchase intention. In addition, construct conceptualizations and operationalizations change over time (Bergkvist & Eisend, 2021), while the theoretical entity of interest (i.e., the conceptual variable) generally remains the same. These findings suggest there is no set way to perfectly measure a concept (Viswanathan, 2022). Nevertheless, much progress has been made in more accurately measuring concepts, particularly in consideration of multi-item versus single-item measures, and the addition of improved quantitative metrics for assessing both reliability and validity.

Model Estimation Implications

An important characteristic of PLS-SEM is that the method does not simultaneously compute all the model relationships, but instead uses separate ordinary least squares regressions to estimate the model's partial regression relationships (Exhibit 1.4)— as implied by its name. As a result, the overall number of model parameters can be extremely high in relation to the sample size as long as each partial regression relationship draws on a sufficient number of observations (Chin & Newsted, 1999). For example, Antioco, Moenaert, Feinberg, and Wetzels (2008) estimate a PLS path model comprising 18 constructs and 33 structural model relationships with merely 121 observations. Needless to say, such an approach is unlikely to be reasonable from a sampling theory perspective unless the population of interest is highly homogeneous. Yet, it offers sufficient statistical power.

Upon convergence, the PLS-SEM algorithm produces a single specific (i.e., determinate) score for each observation per construct. Using these scores as input, PLS-SEM applies ordinary least squares regression with the objective of maximizing the explained variance values of the endogenous constructs and their indicators. That is, the method seeks to maximize **explanatory power**, also referred to as **in-sample predictive power**, which refers to the model's ability to reproduce the data that has been used to estimate the model parameters. Yuan and Fang (2022) have shown, numerically and by a simulation study, that PLS-SEM yields higher levels of explanatory power than CB-SEM. That is, when the aim is to maximize explanatory power, researchers should prefer PLS-SEM over CB-SEM.

The PLS-SEM results also enable assessment of the model's out-of-sample predictive power (or simply predictive power), which indicates a model's ability to predict new or future observations. As high explanatory power (R^2) does not guarantee significant predictive power (Inoue & Kilian, 2005; Sarstedt & Danks, 2022; Shmueli, 2010), researchers need to explicitly test this aspect of their model's performance using holdout samples

(Cepeda-Carrión, Henseler, Ringle, & Roldán, 2016), *k*-fold cross-validation (Shmueli, Ray, Velasquez Estrada, & Chatla, 2016; Shmueli et al., 2019) or a stand-alone predictive ability test (Sharma, Liengaard, Hair, Sarstedt, & Ringle, 2023). These characteristics define PLS-SEM's causal-predictive paradigm in which the aim is to assess the predictive power of a specified model carefully developed on the grounds of theory or logic. The underlying causal-predictive logic follows what Gregor (2006) refers to as **explaining and predicting (EP) theories**. EP theories imply an understanding of the underlying causes and prediction as well as a description of theoretical constructs and their relationships. According to Gregor (2006, p. 626), this type of theory "corresponds to commonly held views of theory in both the natural and social sciences." Numerous seminal theories and models, such as Oliver's (1980) expectation-disconfirmation theory or the various technology acceptance models (e.g., Venkatesh, Morris, Davis, & Davis, 2003) follow an EP-theoretic approach in that their aim is to explain *and* predict. PLS-SEM is perfectly suited to investigate models derived from EP theories as the method strikes a balance between machine learning methods, which are fully predictive in nature, and CB-SEM, which focuses on confirmation and model fit (Richter, Cepeda Carrión, Roldán, & Ringle, 2016). Its causal-predictive nature makes PLS-SEM particularly appealing for research in fields that aim to derive recommendations for practice (Chin et al., 2020; Sarstedt & Danks, 2022). For example, recommendations in managerial implications sections that are an element of many business research journals always include predictive statements ("our results suggest that managers should…"). Making such statements requires a prediction focus in model estimation and evaluation. PLS-SEM perfectly emphasizes this need as the method sheds light on the mechanisms (i.e., the structural model relationships) through which the predictions are generated (Hair, 2021; Hair & Sarstedt, 2021; Legate, Hair, Lambert, & Risher, 2021).

ORGANIZATION OF THE REMAINING CHAPTERS

The remaining chapters provide more detailed information on advanced analyses using PLS-SEM, including specific examples of how to use the SmartPLS 4 software. In this book on advanced PLS-SEM issues, we build upon Stage 7 of the systematic procedure for applying PLS-SEM (Exhibit 1.7) from the *Primer on Partial Least Squares Structural Equation Modeling (PLS-SEM)* (Hair, Hult, Ringle, & Sarstedt, 2022), which enjoys an extremely high level of popularity (Ringle, 2019). The advanced PLS-SEM analyses will enable you to better understand and explain your results, and provide the types of analyses and diagnostic metrics editors and reviewers increasingly request. Exhibit 1.8 displays the chapters and topics covered in this book.

Chapter 2 introduces higher-order constructs, which allow measuring a conceptual variable on different levels of abstraction. For this purpose, a higher-order construct simultaneously models several subcomponents that cover more concrete traits of the conceptual variable represented by this construct. With the growing complexity

EXHIBIT 1.7   ■   A Systematic Procedure for Applying PLS-SEM

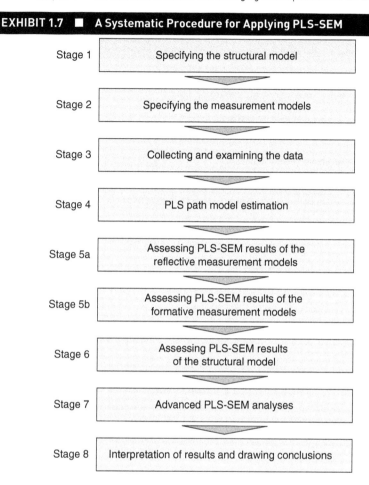

Stage 1	Specifying the structural model
Stage 2	Specifying the measurement models
Stage 3	Collecting and examining the data
Stage 4	PLS path model estimation
Stage 5a	Assessing PLS-SEM results of the reflective measurement models
Stage 5b	Assessing PLS-SEM results of the formative measurement models
Stage 6	Assessing PLS-SEM results of the structural model
Stage 7	Advanced PLS-SEM analyses
Stage 8	Interpretation of results and drawing conclusions

of theories and cause-effect models in the social sciences, researchers have increasingly used these models in their PLS-SEM studies (e.g., Sarstedt, et al., 2022).

Chapter 3 starts with an introduction to nonlinear relationships, which have gained increasing prominence in applications of PLS-SEM (Basco, Hair, Ringle, & Sarstedt, 2021). When the relationship between two constructs is nonlinear, the size of the effect between them depends not only on the magnitude of change in the exogenous construct but also on its value. We introduce the principles of nonlinear modeling and describe how to run corresponding analyses in SmartPLS 4. The second part of the chapter introduces confirmatory tetrad analysis (CTA-PLS), which facilitates empirically assessing whether the data support a formative or a reflective measurement model specification. In light of the potential biases that result from misspecifying measurement models, the CTA-PLS offers a valuable means to safeguard the results' validity.

Chapter 4 introduces two techniques that extend standard PLS-SEM results assessment procedures. We first introduce the necessary condition analysis (NCA) that

EXHIBIT 1.8 ■ Thematical Vverview of this Book	
Chapter	**Topics**
2	Higher-Order Constructs
3	Advanced Modeling and Model Assessment
	— Nonlinear Relationships
	— Confirmatory Tetrad Analysis (CTA-PLS)
4	Advanced Results Illustration
	— Necessary Condition Analysis (NCA)
	— Importance-Performance Map Analysis (IPMA)
5	Modeling Observed Heterogeneity
	— Measurement Invariance Assessment (MICOM)
	— Multigroup Analysis
6	Modeling Unobserved Heterogeneity
	— Finite Mixture PLS (FIMIX-PLS)
	— PLS Prediction-Oriented Segmentation (PLS-POS)

takes a different perspective on the model relationships by testing the degree to which an outcome—or a certain level of an outcome—depends on the values of other constructs or indicators in the model. By assuming such necessity relationships, the NCA complements the sufficiency logic that standard PLS-SEM analyses rely on. In the second part of this chapter, we present the importance performance map analysis (IPMA). The IPMA contrasts the structural model relationships, which represent a construct's or an indicator's "importance" for a target construct with the average construct scores, which represent their "performance" in the PLS path model.

The following two chapters deal with different concepts that enable researchers to model heterogeneous data. Chapter 5 provides an overview of observed and unobserved heterogeneity, showing how disregarding heterogeneous data structures can generate biased results. Next, we discuss measurement invariance, which is a primary concern before comparing groups of sample data. The chapter concludes with an introduction of different types of multigroup analysis that are used to compare parameters (usually path coefficients) between two or more groups of data. While these methods enable researchers to account for observed heterogeneity, more often than not, situations arise in which differences related to unobserved heterogeneity prevent the derivation of accurate results as the analysis on the aggregate data level masks group-specific effects. Chapter 6 introduces two additional methods, finite mixture PLS (FIMIX-PLS) and prediction-oriented segmentation in PLS (PLS-POS) that enable researchers to identify and treat unobserved heterogeneity in PLS path models.

CASE STUDY ILLUSTRATION

Corporate Reputation Model

The most effective way to learn how to use a statistical method is to apply it to a set of data. Throughout this book, we use a single example that enables us to do that. The example is drawn from a series of published studies on corporate reputation, which is general enough to be understood by researchers from various disciplines, thus further facilitating comprehension of the analyses presented in this book. More precisely, we draw on the corporate reputation model by Eberl (2010), which Hair, Hult, Ringle, and Sarstedt (2022) use in their *Primer on Partial Least Squares Structural Equation Modeling (PLS-SEM)*. The model's purpose is to explain the effects of corporate reputation—a company's overall evaluation by its stakeholders (Helm, Eggert, & Garnefeld, 2010)—on customer satisfaction (*CUSA*) and, ultimately, customer loyalty (*CUSL*). Following Schwaiger (2004), corporate reputation is measured using two dimensions. One dimension represents the cognitive evaluations of the company and measures the construct describing the company's competence (*COMP*). The second dimension captures affective judgments and assesses perceptions of the company's likeability (*LIKE*). These two constructs are hypothesized to explain variations in customer satisfaction and loyalty (Schwaiger, Witmaier, Morath, & Hufnagel, 2021). Schwaiger (2004) further identifies four antecedent dimensions of reputation: quality (*QUAL*), performance (*PERF*), attractiveness (*ATTR*), and corporate social responsibility (*CSOR*). Exhibit 1.9 shows the corporate reputation model.

The measurement models of the *LIKE*, *COMP*, and *CUSL* constructs have three reflective indicators, whereas *CUSA* is measured with a single item. In general, we

EXHIBIT 1.9 ■ Structural Model of Corporate Reputation

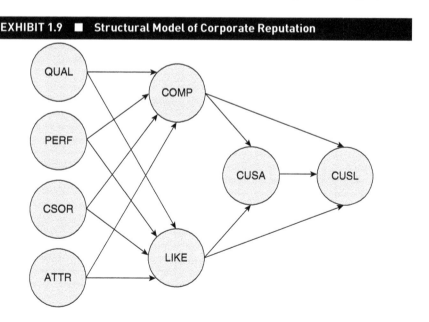

recommend that using single items should be avoided, particularly in PLS-SEM analyses (e.g., Diamantopoulos, Sarstedt, Fuchs, Kaiser, & Wilczynski, 2012; Sarstedt, Diamantopoulos, & Salzberger, 2016; Sarstedt, Diamantopoulos, Salzberger, & Baumgartner, 2016). This single-item measure was included in our example, however, for illustrative purposes of how such a question could be included in obtaining a PLS-SEM solution. Finally, the four exogenous constructs are measured by a total of 21 formative indicators. Exhibit 1.10 provides an overview of all items and item wordings. Respondents rated the questions on 7-point Likert scales, with higher scores denoting higher levels of agreement with a particular statement. In the case of the *CUSA* indicator, higher scores denote higher levels of satisfaction (1 = *very dissatisfied;* 7 = *very satisfied*). Satisfaction and loyalty were measured with respect to respondents' own mobile phone service providers.

The measurement approach has been validated in different countries and applied in various research studies (e.g., Eberl & Schwaiger, 2005; Hult, Hair, Proksch, Ringle, Sarstedt, & Pinkwart, 2018; Radomir & Wilson, 2018; Raithel & Schwaiger, 2014; Raithel, Wilczynski, Schloderer, & Schwaiger, 2010; Schwaiger, Witmaier, Morath,

---

EXHIBIT 1.10 ■ Overview of Constructs and Indicators

Competence (*COMP*)

comp_1	[The company] is a top competitor in its market.
comp_2	As far as I know, [the company] is recognized worldwide.
comp_3	I believe that [the company] performs at a premium level.

Likeability (*LIKE*)

like_1	[The company] is a company that I can better identify with than other companies.
like_2	[The company] is a company that I would regret more not having if it no longer existed than I would other companies.
like_3	I regard [the company] as a likable company.

Customer Loyalty (*CUSL*)

cusl_1	I would recommend [the company] to friends and relatives.
cusl_2	If I had to choose again, I would choose [the company].
cusl_3	I will remain a customer of [the company] in the future.

Customer Satisfaction (*CUSA*)

cusa	If you consider your experiences with [the company], how satisfied are you with [the company]?

Quality (*QUAL*)

qual_1	The products/services offered by [the company] are of high quality.

EXHIBIT 1.10 ■ Overview of Constructs and Indicators (Continued)

qual_2	[The company] is an innovator, rather than an imitator with respect to [the industry].
qual_3	[The company]'s products/services offer good value for money.
qual_4	The services [the company] offered are good.
qual_5	Customer concerns are held in high regard at [the company].
qual_6	[The company] is a reliable partner for customers.
qual_7	[The company] is a trustworthy company.
qual_8	I have a lot of respect for [the company].

Performance (*PERF*)

perf_1	[The company] is a very well-managed company.
perf_2	[The company] is an economically stable company.
perf_3	The business risk for [the company] is modest compared to its competitors.
perf_4	[The company] has growth potential.
perf_5	[The company] has a clear vision about the future of the company.

Corporate Social Responsibility (*CSOR*)

csor_1	[The company] behaves in a socially conscious way.
csor_2	[The company] is forthright in giving information to the public.
csor_3	[The company] has a fair attitude toward competitors.
csor_4	[The company] is concerned about the preservation of the environment.
csor_5	[The company] is not only concerned about profits.

Attractiveness (*ATTR*)

attr_1	[The company] is successful in attracting high-quality employees.
attr_2	I could see myself working at [the company].
attr_3	I like the physical appearance of [the company] (company logo, buildings, shops, etc.).

& Hufnagel, 2021; Schloderer, Sarstedt, & Ringle, 2014; Sharma, Shmueli, Sarstedt, Danks, & Ray, 2021). Research has shown that, compared to alternative reputation measures, Schwaiger's (2004) approach performs favorably in terms of convergent validity and predictive validity (Sarstedt, Wilczynski, & Melewar, 2013). The data set used for all analyses in this book stems from Hair, Hult, Ringle, and Sarstedt (2022) and has 344 responses regarding four major mobile network providers in Germany's mobile communications market (for a newer dataset, see also Sarstedt, Ringle, & Iuklanov, 2023).

PLS-SEM Software

To establish and estimate PLS path models, users can choose from a range of software programs. A popular early example of a PLS-SEM software program is PLS-Graph (Chin, 1994), which is a graphical interface to Lohmöller's (1987) LVPLS, the first user-friendly PLS software. Compared to the original LVPLS, which required the user to enter commands via a text editor, PLS-Graph represents a significant improvement, especially in terms of user-friendliness. With the increasing dissemination of PLS-SEM in a variety of disciplines, several other programs with user-friendly graphical interfaces were introduced to the market, such as SmartPLS (Ringle, Wende, & Becker, 2022) and WarpPLS (Kock, 2020). Finally, users with experience in the statistical software environment R can also draw on packages, such as csem (Rademaker et al., 2021) and SEMinR (Ray et al., 2022), which facilitate the flexible analysis of PLS path models. The new R software workbook of the primer on PLS-SEM (Hair et al., 2022), with an electronic copy download available for free, illustrates all elements of the corporate reputation case study using the SEMinR package.

To date, SmartPLS is the most comprehensive and advanced program in the field (Sarstedt & Cheah, 2019). The software's most recent version, SmartPLS 4, therefore serves as the basis for all case study examples in this book. The student version of the software is available free of charge at https://www.smartpls.com. It offers practically all functionalities of the full version but is restricted to data sets with a maximum of 100 observations. However, as the data set used in this book has more than 100 observations (344 to be precise), we use the professional version of SmartPLS 4, which is available as a 30-day trial version at https://www.smartpls.com. The SmartPLS website includes many additional resources, such as short explanations of PLS-SEM and software-related topics, a list of recommended literature, answers to frequently asked questions, tutorial videos for getting started using the software, and the SmartPLS forum, which enables you to discuss PLS-SEM topics and share ideas with other users.

Setting Up the Model in SmartPLS

Before we specify our model in SmartPLS 4, we need to have data that serve as the basis for running the model. SmartPLS 4 supports data imported from various file formats, such as Microsoft Excel (.xlxs), SPSS (.sav), comma-separated values (.csv), and text (.txt). The only aspect we have to pay attention to is that the first data row contains the variable names in text format and otherwise only numerical values (no text or special characters; also no numerical values in scientific format, e.g., 10E-7). The data we use with the reputation model can be downloaded either as a comma-separated value (.csv) or text (.txt) data set in the download section at https://www.pls-sem.com. Click on **Save Target As . . .** to save the data to a folder on our hard drive or cloud drive. Next, run the SmartPLS software. When we use the **New project** option in the toolbar, a new project will be created. Then, we can use the **Import data file** option in the newly created project. Alternatively, we can right-click on the project and use the **Import data file** option. With this data, as explained in Chapter 2 of the book by Hair, Hult, Ringle, and Sarstedt (2022), we can create the PLS path model as shown in Exhibit 1.11.

EXHIBIT 1.11 ■ Corporate Reputation Model in SmartPLS

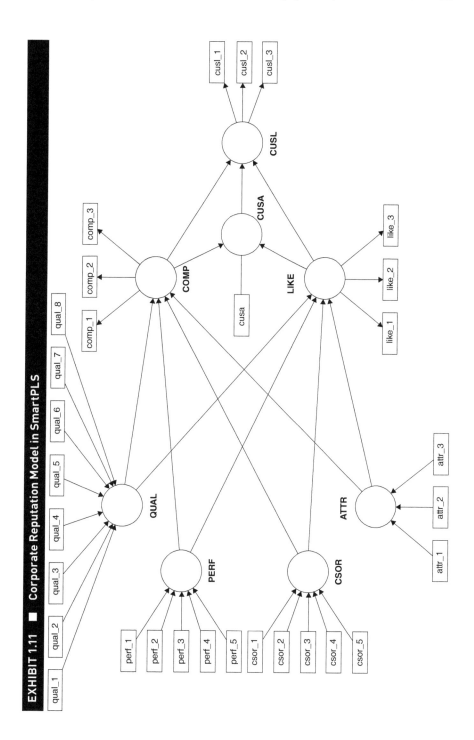

Alternatively, after starting the SmartPLS software, various sample projects appear in the main window under **Sample Projects**, which can be installed directly. When clicking on the **Install** link next to **Example – corporate reputation (advanced)**, the advanced corporate reputation project will appear in the **Workspace** window (Exhibit 1.12), located at the left of the screen. Next, double-click on the **Corporate reputation model**. Then the PLS path model as shown in Exhibit 1.11 will appear in the SmartPLS modeling window.

Following the systematic procedure for applying PLS-SEM presented in Hair, Hult, Ringle, and Sarstedt (2022), the next steps entail the evaluation of the reflectively and formatively specified measurement models, followed by an assessment of the structural model. Readers are advised to consult the *Primer on Partial Least Squares Structural Equation Modeling (PLS-SEM)* (Hair, Hult, Ringle, & Sarstedt, 2022) for a detailed discussion and illustration of these analysis steps. The case study illustrations

EXHIBIT 1.12 ■ The SmartPLS Workspace

in the following chapters will depart from here, assuming that the quality of the original model's measurement and structural models have been established.

SUMMARY

- **Understand the origins and evolution of PLS-SEM.** The precursors to PLS-SEM were two methods (i.e., principle component regression and PLS-R) that used least squares estimation to develop solutions for single and multicomponent models. Further development of these procedures by Herman Wold led to the NIPALS algorithm and a revised generalized version of the PLS algorithm that focused on finding constructs. In the 1980s Herman Wold proposed his "soft modeling basic design" underlying PLS-SEM as an alternative to CB-SEM. The latter method has been labeled as "hard" modeling due to its more stringent assumptions in terms of data distribution and sample size compared to PLS-SEM. While both approaches were developed at about the same time, CB-SEM became much more widely applied because of its early availability through the LISREL software in the late 1970s. It was not until the debut of Wynne Chin's PLS-Graph software in the mid-1990s that PLS-SEM applications began increasing. With the release of SmartPLS 2 in 2005, PLS-SEM's usage grew exponentially.

- **Comprehend model specification in a PLS-SEM framework.** The model specification in PLS-SEM involves two sub-models—the structural model and the measurement model. Whereas the structural model specifies the relationships between the constructs, the measurement models express how to measure the construct utilizing a set of indicators. Measurement models can be specified reflectively, using effect (i.e., reflective) indicators, or formatively, using causal or composite indicators. Whereas constructs measured with causal indicators have an error term, this is not the case with composite indicators, which define the construct in full. Traditionally, composite indicators have been viewed as a means to combine several variables to represent some new entity, whose meaning is defined by the choice of indicators. However, more recent research argues that composite indicators can be used to measure any type of property to which the focal concept refers, including attitudes, perceptions, and behavioral intentions.

- **Describe the PLS-SEM algorithm's basic functioning principles.** The PLS-SEM algorithm uses the empirical data for the indicators and iteratively determines the construct scores, the path coefficients, indicator loadings and weights, and further statistics, such as R^2 values and measures of the model's out-of-sample predictive power. After initialization, the algorithm estimates structural and measurement model parameters seperately, holding the other model elements constant in each iteration. The algorithm's goal is to estimate parameters so residuals in the structural and measurement models are minimized. The results are typically standardized, meaning that, for example, the sizes of the path coefficients can be directly compared with each other even when the model estimation draws on differently scaled indicators. Recent extensions of the original PLS-SEM algorithm facilitate including sampling weights or estimating solutions comparable to common factor models.

- **Understand PLS-SEM's key characteristics vis-à-vis CB-SEM.** A crucial conceptual difference between PLS-SEM and CB-SEM relates to the way each method treats the constructs included in the model. CB-SEM considers the constructs as common factors, whereas PLS-SEM follows a composite model perspective using weighted composites of indicator variables to calculate scores that represent the constructs. Estimating a common factor model using PLS-SEM produces a bias, but the bias produced by CB-SEM when estimating composite models is much more substantial. In model estimation, PLS-SEM follows a causal-predictive paradigm in that the method seeks to maximize (in-sample) prediction of a specified model developed based on theory and logic. Because of the way PLS-SEM estimates the model parameters, the method is not constrained by identification issues, even if the model becomes complex and is being estimated with little data—a situation that typically restricts CB-SEM's use.

REVIEW QUESTIONS

1. Who developed the PLS-SEM algorithm and what was the intention behind its development?

2. What is the difference between common factor models and composite models?

3. What is the difference between reflective and formative measurement models?

4. How does the PLS-SEM algorithm work?

5. What are the key characteristics that distinguish PLS-SEM from CB-SEM?

CRITICAL THINKING QUESTIONS

1. Under what condition is PLS-SEM the preferred method over CB-SEM for prediction, and why?

2. Please comment on the following statement: "Indicators in formative measurement models are error-free."

3. What is the difference between causal and composite indicators?

4. Are common factor models and reflective measurement models the same?

5. Why should researchers test the predictive power of their PLS path models?

KEY TERMS

Artifacts

Causal indicators

Common factor models

Common factor-based SEM

Common variance

Composite indicators

Composite-based SEM

Consistent PLS-SEM (PLSc-SEM)

Correlation weights

Covariance-based structural equation modeling (CB-SEM)

Effect indicators

Endogenous constructs

Error terms

Exogenous constructs

Explaining and predicting (EP) theories

Explanatory power

Factor (score) indeterminacy

Factor weighting scheme

Formative measurement model

In-sample predictive power

Inner model

Measurement models

Metrological uncertainty

Mode A

Mode B

Outer models

Partial least squares path modeling (PLS-SEM)

Partial least squares regression (PLS-R)

Path weighting scheme

PLS-SEM algorithm

PLS-SEM bias

Principal components regression

Reflective indicators

Reflective measurement model

Regression weights

Reliability coefficient ρ_A

Sampling weights

Single-item measurement

Structural model

Sum scores

Weighted PLS-SEM (WPLS)

SUGGESTED READINGS

Bollen, K. A. (2011). Evaluating effect, composite, and causal indicators in structural equation models. *MIS Quarterly*, *35*(2), 359–372.

Bollen, K. A., & Diamantopoulos, A. (2017). In defense of causal-formative indicators: A minority report. *Psychological Methods*, *22*(3), 581–596.

Chin, W. W., Cheah, J-H., Liu, Y., Ting, H., Lim, X.-J., & Cham, T. H. (2020). Demystifying the role of causal-predictive modeling using partial least squares structural equation modeling in information systems research. *Industrial Management & Data Systems*, *120*(12), 2161–2209.

Cook, D. R., & Forzani, L. (2023). On the role of partial least squares in path analysis for the social sciences. *Journal of Business Research*, *167*, 114132.

Hair, J. F., Hult, G. T. M., Ringle, C. M., & Sarstedt, M. (2022). *A primer on partial least squares structural equation modeling (PLS-SEM)* (3rd ed.). Sage.

Hair, J. F., Hult, G. T. M., Ringle, C. M., Sarstedt, M., & Thiele, K. O. (2017). Mirror, mirror on the wall: A comparative evaluation of composite-based structural equation modeling methods. *Journal of the Academy of Marketing Science, 45*(5), 616–632.

Hair, J. F., Hult T., Ringle C. M., Sarstedt, M., Danks, N., & Ray, S. (2022). *Partial least squares structural equation modeling (PLS-SEM) using R: A workbook.* Springer.

Hair, J. F., Sarstedt, M., & Ringle, C. M. (2019). Rethinking some of the rethinking of partial least squares. *European Journal of Marketing, 53*(4), 566–584.

Rigdon, E. E. (2012). Rethinking partial least squares path modeling: In praise of simple methods. *Long Range Planning, 45*(5–6), 341–358.

Rigdon, E. E., Becker, J.-M., & Sarstedt, M. (2019). Factor indeterminacy as metrological uncertainty: Implications for advancing psychological measurement. *Multivariate Behavioral Research, 54*(3), 429–443.

Rigdon, E. E., & Sarstedt, M. (2022). Accounting for uncertainty in the measurement of unobservable marketing phenomena. In H. Baumgartner & B. Weijters (Eds.), *Review of Marketing Research*, vol. 19 (pp. 53–73). Emerald.

Rigdon, E. E., Sarstedt, M., & Ringle, C. M. (2017). On comparing results from CB-SEM and PLS-SEM: Five perspectives and five recommendations. *Marketing ZFP, 39*(3), 4–16.

Ringle, C. M., Sarstedt, M., Sinkovics, N., & Sinkovics, R. R. (2023). A perspective on using partial least squares structural equation modelling in data articles. *Data in Brief, 48*, 109074.

Sarstedt, M., Hair, J. F., Ringle, C. M., Thiele, K. O., & Gudergan, S. P. (2016). Measurement issues with PLS and CB-SEM: Where the bias lies! *Journal of Business Research, 69*(10), 3998–4010.

Sarstedt, M., Ringle, C. M., & Hair, J. F. (2023). PLS-SEM: Indeed a silver bullet–retrospective observations and recent advances. *Journal of Marketing Theory & Practice, 31*(3), 261–275.

Schuberth, F., Müller, T., & Henseler, J. (2021). Which equations? An inquiry into the equations in partial least squares structural equation modeling. In I. Kemény & Z. Kun (Eds.), *New perspectives in serving customers, patients, and organizations: A Festschrift for Judit Simon* (pp. 96–115). Corvinus University of Budapest.

Yuan, K.-H., & Fang, Y. (2022). Which method delivers greater signal-to-noise ratio: Structural equation modelling or regression analysis with weighted composites? *British Journal of Mathematical and Statistical Psychology.* Advance online publication.

2

HIGHER-ORDER CONSTRUCTS

LEARNING OUTCOMES

1. Understand the logic and usefulness of higher-order constructs.

2. Appreciate the different types of higher-order constructs and understand how to specify them in PLS-SEM.

3. Comprehend how to estimate higher-order constructs using the SmartPLS software as well as how to interpret the results.

CHAPTER PREVIEW

With the rising complexity of theories and cause-effect models in the social sciences, researchers have increasingly used higher-order constructs in their PLS-SEM studies (e.g., Sarstedt, Hair, Pick, Liengaard et al., 2022). Higher-order constructs differ from regular constructs in that they include a more general component that measures a conceptual variable at a higher level of abstraction, while simultaneously including several subcomponents, each of which measure more concrete traits of the concepts. Higher-order constructs permit reducing the number of structural model relationships, making the PLS path model more parsimonious, while increasing the bandwidth of content covered by certain constructs (e.g., Johnson, Rosen, & Chang, 2011), and facilitating minimization of multicollinearity. In this chapter, we describe the nature of higher-order constructs and discuss how to develop and validate them in a PLS-SEM context.

HIGHER-ORDER CONSTRUCTS

Terminology and Motivation

Most PLS path models like those covered in the *Primer on Partial Least Squares Structural Equation Modeling (PLS-SEM)* (Hair, Hult, Ringle, & Sarstedt, 2022) deal with first-order constructs. These constructs represent conceptual variables, such as customer engagement, satisfaction, or loyalty using a set of items that capture a single layer of abstraction. In some instances, however, the constructs researchers wish to examine are quite complex and can also be operationalized at higher levels of abstraction. Establishing a **higher-order model** or a **hierarchical component model**, as they are sometimes referred to in the context of PLS-SEM (Lohmöller, 1989, Chapter 3; Wold, 1982), most often involves testing a **second-order construct** that contains two layered structures of constructs. For example, satisfaction can be measured at two levels of abstraction. An ensuing **higher-order construct** would include a general satisfaction construct along with several subconstructs that capture different more concrete attributes of satisfaction, such as satisfaction with the price, satisfaction with the service quality, satisfaction with the personnel, or satisfaction with the servicescape. These more concrete lower-order attributes then form the more abstract, higher-order satisfaction construct, as shown in Exhibit 2.1.

Instead of modeling the attributes of satisfaction as drivers of the respondent's overall satisfaction—or other target constructs (e.g., customer loyalty)—higher-order modeling involves simultaneously mapping the **lower-order components (LOCs)** and a single **higher-order component (HOC)**. Theoretically, this process can be extended to any number of layers yielding third-, fourth-, or fifth-order models, but researchers usually restrict their modeling to two layers (i.e., second-order models).

There are several reasons for including higher-order constructs in a PLS path model (e.g., Edwards, 2001; Johnson, Rosen, & Chang, 2011; Polites, Roberts, & Thatcher, 2012). One reason pertains to the **bandwidth-fidelity tradeoff**, or the idea that broader constructs are better predictors of criteria that span over multiple domains and/or periods of time. That is, if the goal is to predict broadly defined behaviors, then higher-order constructs might prove valuable. Another reason is to overcome the

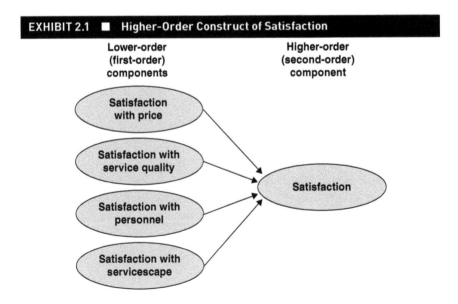

EXHIBIT 2.1 ■ Higher-Order Construct of Satisfaction

jangle fallacy, which occurs when a single phenomenon is examined separately under the guise of two or more variables with different labels.

From a practical perspective, higher-order constructs enable researchers to reduce the number of relationships in the structural model, making the PLS path model more parsimonious, easier to comprehend, and reducing multicollinearity among antecedent constructs. Exhibit 2.2 illustrates this aspect. As can be seen in the complex model, there are nine structural model relationships linking the exogenous constructs (Y_1, Y_2, and Y_3) with the endogenous constructs (Y_4, Y_5, and Y_6). By including a higher-order construct, the number of path coefficients can be reduced to six, yielding a more parsimonious model in terms of structural model relationships. In this case, the HOC is assumed to fully mediate the LOCs' effects on the endogenous constructs (for more detail on mediation, see Hair, Hult, Ringle, & Sarstedt, 2022; Sarstedt, Hair, Nitzl, Ringle, & Howard, 2020). This reduction in model complexity may come at the expense of explanatory power with respect to the endogenous constructs that the HOC explains (i.e., Y_4, Y_5, and Y_6 in Exhibit 2.2). The reason is, different from a direct effects model where all exogenous constructs explain one endogenous construct (Exhibit 2.2, left panel), in the higher-order construct set-up, only the HOC explains the endogenous constructs (Exhibit 2.2, right panel).

Finally, higher-order constructs also prove valuable if formative indicators in a construct's measurement model exhibit high levels of collinearity. High collinearity among the indicators of a formative measurement model can result in biased weights and their signs being reversed. Furthermore, collinearity increases standard errors and, thus, trigger Type II errors (i.e., false negatives; Hair, Hult, Ringle, & Sarstedt, 2022, Chapter 5). Higher-order models facilitate handling of collinearity problems by offering a means to rearrange measurement models. Provided that measurement theory supports this step, researchers can split up the set of indicators and establish separate constructs in a higher-order structure. Consider, for example, a formatively measured

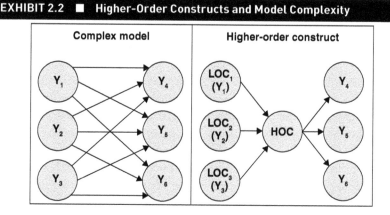

EXHIBIT 2.2 ■ Higher-Order Constructs and Model Complexity

construct with four indicators $(x_1 - x_4)$, of which x_1 and x_2 as well as x_3 and x_4 are highly correlated. If conceptually meaningful, researchers could split up the formative construct into two LOCs, each one being measured with noncollinear indicators (e.g., x_1 and x_3 on the one hand and x_2 and x_4 on the other).

Types of Higher-Order Constructs

Establishing a higher-order structure requires researchers to develop and use an appropriate operational definition of the conceptual variable under consideration. The operational definition facilitates conceptualizing an abstract idea so that it represents the scope of measurable, observable qualities that can be studied (see Chapter 1). The operational definition guides the identification of relevant LOCs, each of which refers to a distinctive element (or component) associated with the HOC, and each of which has a set of indicators that can be specified by the distinctiveness of the element that characterizes the LOC. At the same time, this characterizing distinctiveness should also be sufficiently relevant so only those LOCs that are important for the specific study are captured in a higher-order construct. The operational definition, with its characterizing elements, can vary from study to study since a theoretical concept is not per se determined as multidimensional or unidimensional. Rather, a concept can be specified either way, representing different levels of theoretical abstraction (Bollen, 2011).

Conceptually, higher-order constructs can be established following a bottom-up (i.e., inductive) or top-down (i.e., deductive) approach (e.g., Johnson, Rosen, & Chang, 2011). In the **bottom-up approach**, several constructs are combined into a single, more abstract construct. On the contrary, in the **top-down approach**, a more abstract construct is defined to consist of several components, as is the case in the satisfaction example described above (Exhibit 2.1). Even though frequently used in empirical research to reduce model complexity, we do not recommend simply summarizing information in a more abstract construct.

Establishing a higher-order construct in PLS-SEM always involves a loss of information—at least in principle. The reason is that the direct effects between the LOCs (e.g., $Y_1 - Y_3$ in Exhibit 2.2, left panel) and the criterion constructs (e.g., $Y_4 - Y_6$ in Exhibit 2.2, left panel) are being replaced by three indirect effects via the newly established HOC (Exhibit 2.2, right panel). The PLS-SEM method, however, allows estimating all relationships in a wider nomological net of constructs (i.e., without the HOC) without loss of information. Therefore, when structural theory supports the inclusion of a larger number of constructs, which could be summarized in a HOC, justifying their joint consideration in form of a higher-order construct on the grounds of model parsimony only is not sufficient. Instead, higher-order constructs derived in a top-down manner offer researchers additional insights regarding the effects of different components embedded in a specific construct. Here, the researcher's intention is to determine the effect of such components on other constructs in the model via the HOC.

In addition to using theory to identify inclusion criteria for selecting suitable LOCs, the nature of relations among the LOCs and the HOC must be clarified. A HOC is a general concept that is either represented (in the reflective mode) or constituted (in the formative mode) by its specific components (i.e., the LOCs). If the higher-order construct is reflective, the more general HOC manifests itself in several more specific LOCs. That is, the relationships go from the HOC to its LOCs. This type of model is also referred to as a **spread model** (Lohmöller, 1989, Chapter 3). If the higher-order construct is formative, several specific LOCs represent more concrete components that jointly form the more general HOC (Becker, Klein, & Wetzels, 2012; Edwards, 2001; Wetzels, Odekerken-Schroder, & van Oppen, 2009). That is, the relationships go from the LOCs to the HOC. This model type is also referred to as a **collect model** (Lohmöller, 1989, Chapter 3). The higher-order construct in Exhibit 2.1 has formative relationships going from the LOCs to the HOC, representing each LOC's relative contribution to forming the HOC. However, if operationalized differently, these relationships could also have been modeled in the opposite direction with the LOCs reflecting the HOC.

A formative specification is appropriate when the operational definition of the conceptual variable suggests that a change in a LOC's value due to, for example, a change in a respondent's assessment of the trait being captured by the LOC changes the value of the HOC. Analogous to indicators in formative measurement models, the LOCs do not need to, but can, be correlated as they do not represent concrete manifestations of the HOC. In contrast, a reflective specification is appropriate when there is a more general, abstract construct that explains the correlations between the LOCs as shown in Exhibit 2.2. Hence, there should be substantial correlations between the LOCs that—analogous to reflective measurement models in first-order constructs—are assumed to be "caused" by the HOC.

In addition to the measurement specification of the higher-order construct as a whole, as represented by the relationships between the HOC and the LOCs, higher-order constructs need to be characterized on the grounds of the specification of the LOCs' measurement models. The LOCs' measurement models can also be reflective or formative. As a result, four main types of higher-order constructs are possible (Exhibit 2.3), as discussed in the extant literature (Jarvis, MacKenzie, & Podsakoff, 2003; Wetzels, Odekerken-Schroder, & van Oppen, 2009). The **reflective-reflective higher-order construct** (**Type I higher-order construct**) shown in Exhibit 2.3 has reflective relationships between the HOC and the LOCs, and also between the LOCs and the indicators. In this type of higher-order construct, the LOCs are highly correlated and the HOC represents the cause explaining these correlations. Lohmöller (1989, Chapter 3) calls this type of higher-order construct **hierarchical common factor model**, in which the general HOC represents the common factor of several specific factors (i.e., LOCs). The use of reflective-reflective higher-order constructs has been subject to considerable debate, with critics arguing that such models do not exist (or are meaningless)

EXHIBIT 2.3 ■ Types of Higher-Order Constructs

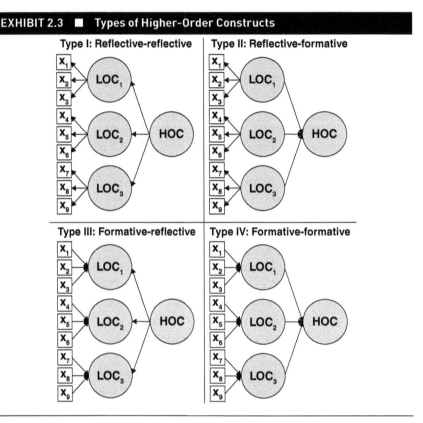

Source: Adapted from Figure B1 in Ringle, Sarstedt, & Straub (2012).

since reflective measures should be unidimensional and conceptually interchangeable, which conflicts with the view of multiple underlying dimensions being distinct in nature (Lee & Cadogan, 2013). That is, if the indicators of each LOC would correlate highly, any indicator should also relate to any other LOC. This would make the LOC-level redundant, implying that the indicators should be directly linked to the primary source of reflection—that is, the HOC (Mikulic, 2022). Temme and Diamantopoulos (2016) bemoan that this line of reasoning rests on the flawed assumption that unidimensionality of the higher-order models' elements (i.e., the LOCs and the HOC) is a necessary condition for reflective measurement (Bollen & Lennox, 1991). However, psychometric theory has long established that indicators can serve as measurements of more than one construct (e.g., Bollen 1989)—as is the case in, for example, bifactor models (e.g., Zhang, Sun, Cao, & Drasgow, 2021). Hence, the assumption that highly correlated indicators in the LOCs' measurement models imply high indicator correlations with all other LOCs stands on quicksand.

Reflective-reflective higher-order constructs might also be used in other settings, for example, in a situation where the LOCs represent different measurements of a concept at different points in time (i.e., different batteries of a test sequence), which the HOC explains simultaneously. Lohmöller (1989, Chapter 3) characterizes this constellation as a **multiple battery model** and presents the example of the general ability of schoolchildren. In this example, each LOC represents a test battery of verbal, numerical, and spatial indicators measured at different points in time (e.g., tests at the beginning, middle, and end of the school year) for the same class (i.e., the same individuals). However, extant reviews of PLS-SEM's use in different fields have not disclosed any applications of the multiple battery model thus far.

In a **reflective-formative higher-order construct** (**Type II higher-order construct**; Exhibit 2.3), the HOC is a combination of several reflectively measured LOCs. That is, the specific LOCs do not necessarily share a common cause but rather form the general HOC—the relationships go from the LOCs to the HOC. Barroso and Picón (2012) offer an example of a reflective-formative higher-order construct in their analysis of perceived switching costs. They identify a set of six dimensions (benefit loss costs, personal relationship loss costs, economic risks costs, evaluation costs, set-up costs, and monetary loss costs) that represent LOCs of the more general HOC, perceived switching costs. Commenting on their measurement specification, Barroso and Picón (2012, p. 532) note: "A modification in one dimension does not necessarily imply a modification in another. In other words, they do not necessarily covary; rather, each dimension can vary independently of the others." For this reason, unlike some prior research, Barroso and Picón (2012) propose that perceived switching costs is an aggregate construct that is expressed as a composition of its different LOCs—see Becker, Klein, and Wetzels (2012) for further examples of reflective-formative higher-order constructs. When the higher-order construct is specified formatively (i.e., the relationships go from the LOCs to the HOC), the HOC fully mediates the relationships between the LOCs and any other construct the higher-order construct explains. For example, in the formative higher-order construct in Exhibit 2.2, all relations from Y_1, Y_2, and Y_3 to the criterion constructs Y_4, Y_5, and Y_6 go through the HOC.

Another higher-order construct is the formative-reflective type (**Type III higher-order construct**; Exhibit 2.3). The **formative-reflective higher-order construct** includes a more general HOC that explains the formatively measured LOCs. The objective of this type is extracting the common part of several formatively measured LOCs that have been established to represent the same theoretical content. However, every LOC builds on a set of different indicators. By using several formatively measured LOCs, researchers can overcome the problem that a stand-alone construct measured with formative indicators can hardly cover the construct's domain in full. Using similar yet distinct formatively measured LOCs as representations of the HOC offers a broader coverage of the construct domain (Becker, Klein, & Wetzels, 2012). A typical example of this higher-order construct type is overall firm performance in which several formatively measured LOCs represent performance-relevant characteristics (e.g., market share, number of employees, or turnover). The HOC represents the common part of the LOCs (i.e., overall firm performance; Jarvis, MacKenzie, & Podsakoff, 2003; Petter, Straub, & Rai, 2007). Alternatively, the formative-reflective higher-order construct can serve as a multiple battery model as explained in the context of the reflective-reflective higher-order construct. In that case, the LOCs represent the same construct that has been formatively measured with the same indicators and for the same observations at different points in time. In this type of multiple battery model, the relationships from the HOC to the LOCs will be of similar magnitude since they represent one concept measured at different points in time.

Finally, the **formative-formative higher-order construct (Type IV higher-order construct**; Exhibit 2.3) determines the relative contribution of the formatively measured LOCs to the more abstract HOC. This type is useful to structure a complex formative construct with many indicators into several subconstructs, as is the case when researchers subsume several concrete aspects under a more general concept. Again, firm performance would represent a concept of this nature that could be measured using this higher-order construct type. While the formative-reflective higher-order construct type would comprise different indices of overall firm performance by the LOCs, the formative-formative type includes LOCs representing different aspects of performance, such as the performance of different organizational activities or subdivisions (e.g., R&D performance, HR performance, sales performance) that together determine overall firm performance (i.e., the HOC; Jarvis, MacKenzie, & Podsakoff, 2003; Petter, Straub, & Rai, 2007), but do not necessarily have to correlate with each other.

Theoretical models with higher-order constructs feature prominently in applications of PLS-SEM as evidenced in various review articles. Sarstedt, Hair, Pick, Liengaard et al.'s (2022) analysis of higher-order construct applications in the top 30 marketing journals between 2011 and 2020 showed that 71 of the 239 analyzed studies (29.71%) included at least one such construct. The majority of these studies proposed second-order constructs (65 studies), while the remaining studies included third-order constructs (5 studies) or both (1 study). Analyzing the construct types used, the authors found that most of the studies employ Type I (reflective-reflective; 30 studies), Type II (reflective-formative; 26 studies), or both (4 studies). Only five studies employ Type IV (formative-formative), while no study draws on a Type III (formative-reflective) measurement specification. In an earlier review, Ringle, Sarstedt, and Straub (2012) found a similar share of higher-order construct types published in the management information systems flagship journal *MIS Quarterly*. Higher-order constructs are typically embedded in a larger nomological network of constructs, in which they serve as an antecedent, consequence, or both. For example, revisiting Sarstedt, Hair, Pick, Liengaard et al.'s (2022) analysis shows that in 67 of the 71 studies (94.37%), the higher-order construct is part of a larger network of constructs. In Ringle, Sarstedt, and Straub (2012), this share was only marginally smaller (86.67%). Therefore, the discussion should not be limited to higher-order constructs as separate constructs, also referred to as a **stand-alone higher-order construct**, but should also consider their potential application in a nomological network of constructs embedded in a structural model (Becker, Klein, & Wetzels, 2012).

Specifying Higher-Order Constructs

Overview

PLS-SEM allows the specification and estimation of all higher-order construct types, as shown in Exhibit 2.3. However, the specific type of HOC demands careful consideration when specifying and estimating the model, since PLS-SEM requires each construct in the PLS path model to have at least one indicator in its measurement model. This necessity holds not only for LOCs but also for the HOC, which is an abstract representation of the conceptual variable under consideration. As such, the nature of a higher-order construct in PLS-SEM is different from that in covariance-based SEM (CB-SEM), where the HOC has no indicators in its measurement model. For this reason, higher-order constructs are sometimes called phantom variables in CB-SEM.

To handle the measurement issue of higher-order constructs in PLS-SEM, researchers can draw on four approaches. In the **repeated indicators approach**, all indicators of the LOCs are assigned to the measurement model of the HOC (Lohmöller, 1989, Chapter 3; Wold, 1982). In the second-order model examples in Exhibit 2.3, the repeated indicators approach would use the indicators x_1 to x_9 of the LOCs to establish the measurement model of the HOC. Consequently, the indicators are used twice: once for the LOCs and again for the HOC. Becker, Klein, and Wetzels (2012) introduced the **extended repeated indicators approach** to handle problems that emerge when the HOC in a reflective-formative or formative-formative higher-order construct has one or more antecedent construct(s). In this case, the LOCs fully explain the variance of the HOC, producing an $R^2 \approx 1.0$. However, with all the HOC's variance being explained by its LOCs, there is no variance left to be explained by an antecedent construct. As a consequence, any path relationship from an antecedent construct to the higher-order construct (i.e., to the HOC) will be close to zero and nonsignificant by design.

Finally, researchers have proposed the embedded (Agarwal & Karahanna, 2000) and disjoint (Wilson, 2010) two-stage approaches, which use construct scores generated in the first stage as input for the model computation in the second stage. Since all approaches generally yield similar results (Cheah, Ting, Ramayah, Memon et al., 2019), there is often no compelling reason to prefer one over the other. However, as the two-stage approach resolves problems that occur in specific model constellations when using the (extended) repeated indicators approach and because of their simple implementation in the SmartPLS 4 software, Becker, Cheah, Gholamzade, Ringle et al. (2023) recommend routinely drawing on the two-stage approaches. We, therefore, focus our discussions on the two-stage approaches (i.e., embedded and disjoint) while mentioning the (extended) repeated indicators' approach where needed.

The Embedded Two-Stage Approach

The first stage of the **embedded two-stage approach** (Agarwal & Karahanna, 2000) corresponds to the standard repeated indicators approach in that all indicators of the LOCs are assigned to the HOC's measurement model. Exhibit 2.4 shows such a set-up in a model where a Type II higher-order construct with three LOCs ($Y_1 - Y_3$) is embedded in a path model with one antecedent construct (Y_6) and one criterion construct (Y_5). As can be seen, all indicators of the LOCs ($x_1 - x_9$) are repeated in the measurement model of the HOC (Y_4). Stage 1 of the embedded two-stage approach entails estimating the model in Exhibit 2.4 (left panel), but instead of interpreting the estimates, researchers need to save the construct scores that result from this analysis.

In Stage 2, these construct scores are used as single-item indicators in the HOC's measurement model as shown in the right panel of Exhibit 2.4. That is, the HOC Y_4 is measured with three formative indicators capturing the construct scores of Y_1, Y_2, and Y_3 from Stage 1. Except for the HOC, all other constructs in the model (e.g., Y_5 in Exhibit 2.4) are measured with single items that capture each construct's scores from the previous stage. Importantly, when using the embedded two-stage approach, the entire path model must already be considered in Stage 1. It is not permissible to estimate the higher-order construct as a stand-alone construct in the first stage of the embedded two-stage approach and successively integrate it into the full path model.

EXHIBIT 2.4 ■ The Embedded Two-Stage Approach

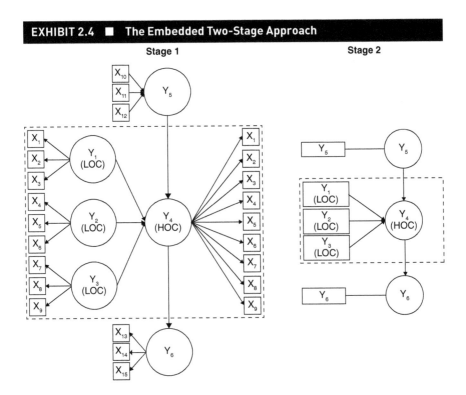

The Disjoint Two-Stage Approach

The **disjoint two-stage approach** (Wilson, 2010) differs from the embedded two-stage approach in the specification of both stages. Rather than using the repeated indicators approach in Stage 1, the disjoint two-stage approach considers only the LOCs of the higher-order construct (i.e., without the HOC) in the path model. These are directly linked to all other constructs that the higher-order construct is theoretically related to (Exhibit 2.5, left panel). We then need to only save the construct scores of the LOCs (e.g., the scores of the constructs Y_1, Y_2, and Y_3 similar to the previous example).

In Stage 2, these scores are then used to establish the HOC's measurement model. However, different from the embedded two-stage approach, all other constructs in the path model are estimated using their standard multi-item measures as in Stage 1 (Exhibit 2.5, right panel).

Recommendations

The two-stage approaches offer researchers more modeling flexibility than the (extended) repeated indicators approach. For example, they allow assessing the

EXHIBIT 2.5 ■ The Disjoint Two-Stage Approach

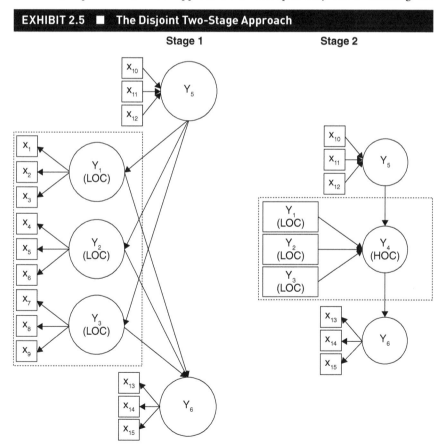

higher-order construct's measurement specification on the grounds of a confirmatory tetrad analysis (CTA-PLS; see Chapter 3). Using the construct scores of the LOCs from Stage 1 as input for the measurement model specification in Stage 2, the CTA-PLS assesses the covariance structure of the HOC's measurement model to test whether the HOC should be specified reflectively or formatively. Such an assessment is not feasible with the repeated indicators approach because the CTA-PLS—as implemented in software programs, such as SmartPLS 4—considers the covariances only in the measurement models and not among the constructs. Based on the CTA-PLS results, researchers gain additional insights regarding the higher-order construct's measurement specification.

While the results of these two approaches do not differ significantly (Cheah, Ting, Ramayah, Memon, Cham, & Ciavolino, 2019), we recommend using the disjoint two-stage approach, as it allows for validating the entire path model on the grounds of the original construct measures in Stage 2 (Becker, Cheah, Gholamzade, Ringle et al., 2023). For example, researchers can use the $PLS_{predict}$ procedure (Shmueli, Ray, Velasquez Estrada, & Chatla, 2016; Shmueli, Sarstedt, Hair, Cheah, Ting, & Ringle, 2019) or the cross-validated predictive ability test (Liengaard, Sharma, Hult, Jensen, Sarstedt, Hair, & Ringle, 2021; Sharma, Liengaard, Hair, Sarstedt, & Ringle, 2023) to estimate the model's predictive power on an indicator level. This sets the disjoint approach apart from the embedded two-stage approach, which is restricted to validating the measurement models in a model that includes the HOC with its repeated indicators (Exhibit 2.4, left panel; i.e., this model set-up, however, only serves identification purposes and does not fully correspond to the path model as hypothesized by the researcher). In addition, the estimates of the relationships between the HOC and the LOCs in Stage 1 of the embedded two-stage approach can be adversely impacted by unevenly distributed numbers of indicators in the LOCs. If, for example, in a reflective formative higher-order construct, one LOC is measured with more indicators than the other LOCs, its weight for forming the HOC will be higher due to the repetition of indicators in the HOC's measurement model. The weight estimate can therefore paint a misleading picture of the LOC's relative relevance for forming the HOC. In light of the above, researchers should draw on the disjoint two-stage approach when specifying higher-order constructs.

Estimating Higher-Order Constructs

The PLS-SEM algorithm draws on two modes to estimate the indicator weights that represent each indicator's relative contribution to forming a construct (Chapter 1). In Mode A indicator weight estimates correspond to the bivariate correlations between each indicator and the construct (i.e., correlation weights). In contrast, Mode B computes indicators by regressing each construct on its associated indicators (i.e., regression weights). While researchers typically use Mode A to estimate reflectively specified measurement models and Mode B to estimate formatively specified measurement models (Hair, Hult, Ringle, & Sarstedt, 2022), Becker, Klein, and Wetzels (2012) show that the choice of measurement mode should consider the relationship between the HOC and the LOCs rather than the operationalization of the HOC when applying the repeated indicators approach. Specifically, their simulation study shows that Mode B estimation of the HOC in a reflective-formative type higher-order construct produces the smallest parameter estimation bias. Hence, even though the (repeated) indicators identifying the HOC are specified reflectively, researchers should use Mode B for estimating these repeated indicators on the HOC. In light of these findings, researchers should use Mode A for reflectively specified higher-order constructs (i.e., reflective-reflective and formative-reflective types) and Mode B for formatively specified higher-order constructs (i.e., reflective-formative and formative-formative types) when estimating the model in Stage 1 of the embedded two-stage approach. In contrast, the disjoint two-stage approach should be estimated using the standard settings on both stages; that is, Mode A for reflectively specified measurement models and Mode B for formatively specified measurement models.

Finally, Becker, Klein, and Wetzels (2012) show that the path weighting scheme (Lohmöller, 1989, Chapters 2) to estimate the PLS path model produces the overall best parameter recovery in formatively specified higher-order constructs (i.e., reflective-formative and formative-formative types). Even though more recent research has not extended Becker, Klein, and Wetzels's (2012) study in this regard, we expect their findings generalize to reflectively specified higher-order constructs (i.e., reflective-reflective and formative-reflective types). Hence, we recommend using the path weighting scheme as the default setting when estimating higher-order constructs in PLS-SEM.

Validating Higher-Order Constructs

When validating higher-order constructs, the same model evaluation criteria generally apply as for any PLS-SEM analysis (Hair, Howard, & Nitzl, 2020; Hair, Hult, Ringle, & Sarstedt, 2022; Ramayah, Cheah, Chuah, Ting, & Memon, 2016). However, researchers need to consider two additional measurement models in higher-order constructs for which the evaluation criteria apply: (1) the measurement models of the LOCs, and (2) the measurement model of the higher-order construct as a whole, represented by the relationships between the HOC and its LOCs. In light of our previous recommendations, we focus our discussion of model validation on the embedded and disjoint two-stage approaches. See Sarstedt, Hair, Cheah, Becker, and Ringle (2019) and Hair, Moisescu, Radomir, Ringle et al. (2020) for a discussion of model validation under the (extended) repeated indicators approach.

The procedures and criteria that have been recommended for the measurement models and the structural model also apply to the PLS-SEM results assessment of the two-stage approach (Hair, Howard, & Nitzl, 2020; Hair, Hult, Ringle, & Sarstedt, 2022; Ramayah, Cheah, Chuah, Ting, & Memon, 2016). That is, the results evaluation in Stage 1 considers all measurement models, including those of the LOCs, but with a single exception. Specifically, when applying the embedded two-stage approach the higher-order construct must *not* be evaluated in terms of its (repeated) indicators directly associated with the HOC (x_1-x_9 in Exhibit 2.4). These indicators only ensure that the higher-order construct is identified—they do not represent its actual measurement model. In fact, the repetition of the LOCs' indicators in the measurement model of the HOC automatically violates discriminant validity as evidenced in high HTMT values. Similarly, the set of indicators assigned to the HOC is not unidimensional by design as they stem from LOCs that represent different concepts.

After model estimation in Stage 2, all measurement models need to be assessed again in terms of reliability and validity, even if their reliability and validity have already been established in the Stage 1 analysis. The reason is that the inclusion of the higher-order construct in Stage 2 changes the model set-up, which entails changes in the model estimates. Generally, however, these changes will not result in reliability or validity issues if these have been confirmed in the Stage 1 analysis. More importantly, this assessment also needs to take the newly established higher-order construct into account whose measurement model is defined by the relationships between its indicators, which come in the form of construct scores derived from Stage 1. That is, in cases of Type I and Type III models, the standard evaluation criteria for reflective measurement models need to be applied, while for Type II and Type IV models, the evaluation criteria for formative measurement models need to be applied.

Once the measures' reliability and validity have been established, the structural model results need to be analyzed, drawing on the standard criteria as documented in, for example, Hair, Hult, Ringle, and Sarstedt (2022). Researchers need to pay particular attention to any structural model assessment that involves interpreting the indicators, such as when using PLS$_{predict}$ (Shmueli, Ray, Velasquez-Estrada, & Chatla, 2016; Shmueli, Sarstedt, Hair, Cheah, Ting, & Ringle, 2019), or running an IPMA (Ringle & Sarstedt, 2016). Specifically, in the case of the embedded two-stage approach, corresponding structural model assessments should be carried out in the first stage. The reason is that Stage 2 uses the construct scores of Stage 1 as single items, which renders validation on the grounds of the items meaningless. In contrast, the disjoint two-stage approach uses multiple items in the second stage, which permits the application of all structural model assessment criteria. Hence, when using the disjoint two-stage approach, researchers should assess the structural model on the grounds of the Stage 2 results. As with the measurement model assessment, this evaluation must disregard the HOC.

Finally, situations occur in which researchers are uncertain whether or not to measure a theoretical concept by means of a higher-order construct. To make a decision, Becker, Cheah, Gholamzade, Ringle et al. (2023) suggest comparing models with and without the higher-order construct. For this purpose, they can use model selection criteria that are well known from the regression literature. Sharma, Sarstedt, Shmueli, Kim, and Thiele (2019) and Sharma, Shmueli, Sarstedt, Danks, and Ray (2018) compared the efficacy of various metrics for model comparison tasks and found that the

Bayesian information criterion (BIC) and the criterion suggested by Geweke and Meese (GM) perform well in selecting a parsimonious model that fits the data well and has a good predictive power. As the BIC is easier to compute, extant literature recommends focusing on this criterion (Hair, Hult, Ringle, & Sarstedt, 2022, Chapter 6). If this analysis suggests the model with the higher-order construct produces lower BIC values compared to the model without the higher-order specification, researchers should consider the higher-order construct in their model specification and estimation. In addition, researchers can compute BIC-based Akaike weights, which offer relative weights of evidence in favor of the models under consideration (Danks, Sharma, & Sarstedt, 2020). Alternatively, researchers can compare the model specifications with regard to their predictive power using Liengaard, Sharma, Hult, Jensen, Sarstedt, Hair, & Ringle's (2021) cross-validated predictive ability test (see also Sharma, Liengaard, Hair, Sarstedt, & Ringle, 2023) to empirically justify its consideration.

Rules of Thumb

The previous descriptions have shown that higher-order constructs are a useful means to retain information while making the structural model more parsimonious. At the same time, their specification and particularly their validation requires special care. In Exhibit 2.6, we summarize the rules of thumb to consider when using higher-order constructs.

CASE STUDY ILLUSTRATION

Drawing on the case study model and data presented in Chapter 1, we outline how to create a higher-order construct for the construct corporate reputation. The corporate reputation model focuses on the *COMP* and *LIKE* constructs representing two separate dimensions of corporate reputation (*REPU*). Instead of modeling the distinct

EXHIBIT 2.6 ■ Rules of Thumb for Using Higher-order Constructs	
Aspect	**Rules of Thumb**
Conceptualization	● Closely examine and describe the theoretical and conceptual foundations of the higher-order construct type.
Specification	● To determine the results, the higher-order construct must be embedded in the nomological network of the underlying path model with its predecessor or successor constructs; it is not permissible to estimate the higher-order construct as a stand-alone construct in the first stage of the embedded two-stage approach.
	● Use the two-stage approaches to specify the higher-order construct. The disjoint two-stage approach should be preferred due to its greater flexibility in model validation.

(Continued)

EXHIBIT 2.6 ■ Rules of Thumb for Using Higher-order Constructs (Continued)	
Aspect	**Rules of Thumb**
Estimation	• Use Mode A to estimate the HOC of a reflective-reflective and formative-reflective type higher-order construct in Stage 1 of the embedded two-stage approach.
	• Use Mode B to estimate the HOC of a reflective-formative and formative-formative type higher-order construct in Stage 1 of the embedded two-stage approach.
	• Use the path weighting scheme.
Validation	• Preferably, use the disjoint two-stage approach and validate the measurement models' reliability and validity in both stages, including those of the LOCs.
	• In case of using the embedded two-stage approach, do not consider the HOC's measurement model, operationalized by repeated indicators, in Stage 1. Also, when using the embedded two-stage approach, apply criteria for assessing the model's explanatory and predictive power in Stage 1.
	• Apply the standard structural model assessment criteria and procedures for assessing the relationships between the higher-order construct and any predecessor and successor constructs.
	• Compare the model with the higher-order construct to the model without the higher-order construct (e.g., by selecting the model specification that produces lower BIC values).

impact of the antecedent constructs (i.e., *ATTR, CSOR, PERF*, and *QUAL*) on *COMP* and *LIKE* as well as their effect on the criterion variables (i.e., *CUSA* and *CUSL*) separately, these two constructs could be handled as subdimensions of a more general *REPU* construct. By establishing a second-order construct with *COMP* and *LIKE* as LOCs, the PLS path model becomes more parsimonious. From a measurement theory perspective, *COMP* and *LIKE* determine *REPU* (e.g., Eberl, 2010), therefore inferring a reflective-formative higher-order construct type specification. Since we deal with only two LOCs, the CTA-PLS procedure, which would require at least four LOCs (see Chapter 3 for more detail), cannot provide additional empirical substantiation concerning the direction of the relationship between the LOCs and the HOC. For the empirical illustration, however, we consider *REPU*'s reflective-formative higher-order construct type specification.

To establish the reflective-formative higher-order construct *REPU*, we draw on the disjoint two-stage approach. Stage 1 requires estimating the model with the LOCs (*COMP* and *LIKE*), but without the HOC *REPU*, which will be included in Stage 2—just like in the original corporate reputation model. Navigate to the SmartPLS

Workspace, and double-click on **Corporate reputation model** in the **Example – Corporate reputation (advanced)** project. To estimate this model, click on **Calculate → PLS-SEM algorithm** in the SmartPLS menu. Alternatively, we can left-click on the wheel symbol with the label **Calculate** in the tool bar. Run the PLS-SEM algorithm using the default settings (i.e., path weighting scheme, standardized results, and default initial weights) and make sure to check the box **Open report** in the lower right corner. After clicking the **Start calculation** button, SmartPLS opens the **Results report**. We find that all construct measures meet the required standards in terms of reliability and validity. For example, measures of *LIKE* yield satisfactory levels of convergent validity (AVE = 0.747) and internal consistency reliability (Cronbach's alpha = 0.831; ρ_A = 0.836; ρ_C = 0.899). Similarly, the measures of *COMP* exhibit convergent validity (AVE = 0.688) and internal consistency reliability (Cronbach's alpha = 0.776; ρ_A = 0.786; ρ_C = 0.869). For a detailed overview of the model evaluations of all reflective and formative measurement models, see Chapters 4 and 5 in Hair, Hult, Ringle, and Sarstedt (2022).

On these grounds, we can proceed to Stage 2 of the disjoint two-stage approach, which requires replacing the LOCs with the HOC, the latter of which is measured using the construct scores of *COMP* and *LIKE* from Stage 1. SmartPLS allows to conveniently process the construct scores from a previous algorithm run and include them in a separate dataset that is being added to an existing project. To do so, select the **Create data file** option in the tool bar of the **Results report**. In the menu that opens, select the **Example – Corporate reputation (advanced)** project, the file name (e.g., **HCM disjoint 2nd stage**), and check the boxes next to **Manifest variable scores** (i.e., to include the indicators used in the model in the new dataset), **Latent variable scores** (i.e., to include new variables for the construct scores in the new dataset), and **Other** (i.e., to include all other variables available in the new dataset). Then, left-click on the **Create** button. Click on the orange arrow button labeled **Edit** in the tool bar to return to the **Modeling window**. Next, click on the orange arrow button labeled **Back** to enter the **Workspace** view where the new dataset appears under the selected project. In the **Example – Corporate reputation (advanced)** project, we now see the newly created **HCM disjoint 2nd stage** dataset. Next, right-click on **Corporate reputation model**. In the menu that opens, select the **Duplicate** option. A dialog opens that allows us to enter a name for the copy of the newly added model. Use a self-explaining name such as **HCM reflective-formative 2nd stage**. After pressing the **Create** button, the new model, which is a duplicate of the corporate reputation model shown in **Corporate reputation model**, appears in the project displayed in the **Workspace**. Double-click on the **HCM reflective-formative 2nd stage** model to open it in the modeling window. On the left-hand side, above the list of indicators, left-click on the **Select dataset** button and choose the newly created dataset (i.e., **HCM disjoint 2nd stage**). The **HCM reflective-formative 2nd stage** model already includes the final model with the HOC. In case we want to set up the Stage 2 model ourselves, we need to delete the *COMP* and *LIKE* constructs, and establish a new construct *REPU*, measured formatively by means of the indicators labeled *LV scores – COMP* and *LV* *scores – LIKE*. Note that by default, newly added constructs have a reflective measurement model in SmartPLS. In order to switch to a formative specification, right-click on *REPU* and select the **Invert measurement model** option. Finally, we need to add paths from *ATTR, CSOR, PERF*, and *QUAL* to *REPU* as well as from *REPU* to *CUSA* and *CUSL*. The final model should look like the **HCM reflective-formative 2nd stage** model in Exhibit 2.7. Estimate the model by going to **Calculate → PLS-SEM algorithm** in the SmartPLS tool bar. Using the **Results report** that opens, we can now assess the reflective and formative measurement models. The results of the regular constructs are very similar to those reported in Chapters 4 and 5 in Hair, Hult, Ringle, and Sarstedt (2022). Therefore, in the following, our assessment will focus on the higher-order construct.

The first step in formative measurement model assessment is to establish the higher-order construct's convergent validity by means of a redundancy analysis (Cheah, Sarstedt, Ringle, Ramayah, & Ting, 2018; Hair, Hult, Ringle, & Sarstedt, 2022, Chapter 5). For this purpose, select the project in the **Workspace** and left-click on the **PLS-SEM** button in the tool bar to create a new model. Choose an intuitive name, such as **HCM redundancy analysis REPU**, and left-click on **Save**. An empty modeling window will appear next. On the left-hand side, above the indicators, make sure the **HCM disjoint 2nd stage** dataset has been selected. Next, drag and drop the two *REPU* indicators (*LV scores–COMP* and *LV scores-LIKE*) to the modeling window and assign a meaningful name (e.g., *REPU_F* in which *F* stands for "formative"). Make sure to change the construct's measurement model to formative. Next, establish a construct labeled *REPU_G* in which *G* stands for "global," indicating the construct is measured by means of the global single item *repu_global* ("[company] has a high reputation"). Then, draw a path from *REPU_F* to *REPU_G*. Estimating the model by going to **Calculate → PLS-SEM algorithm** will produce the results shown in Exhibit 2.8. Since the relationship between *REPU_F* and *REPU_G* is above the threshold of 0.7, we find support for the higher-order construct's convergent validity.

To continue the results assessment, we return to the **HCM reflective-formative 2nd stage** model, estimate it again, and check for potential collinearity among the LOCs. To do so, go to **Quality Criteria → Collinearity Statistics (VIF)**. We find that the VIF value of *LV scores–COMP* and *LV scores–LIKE* (1.686) is considerably below the threshold of 3, providing support that collinearity is not a critical issue. The final step requires assessing the higher-order construct's indicator weights in terms of their significance and relevance. Run the bootstrapping procedure with **10,000** subsamples, the **Most important (faster)** results computation option, the **Percentile bootstrap** confidence interval, **Two tailed** testing, and a significance level of **0.05**. In the **Results report** that opens, navigate to **Final results → Outer weights coefficients → Confidence intervals bias corrected**. The results show that both weights are significant as both confidence intervals' lower bounds are clearly larger than zero. We also find that the impact of *LIKE* (0.682) is stronger than the one of *COMP* (0.416), further emphasizing the relevance of reputation's affective dimension.

EXHIBIT 2.7   ■   Reflective-Formative Higher-Order Construct Example (Stage 2 Model)

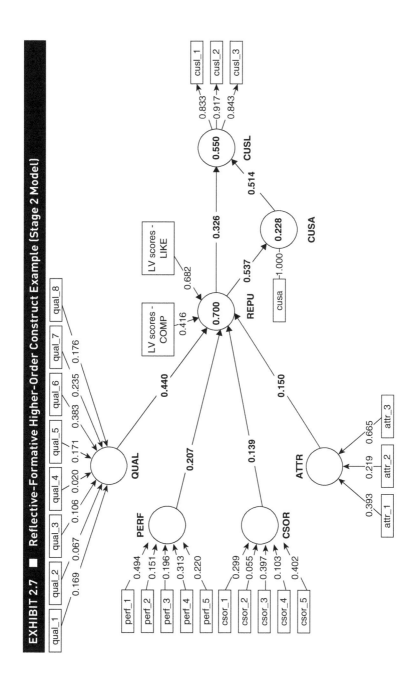

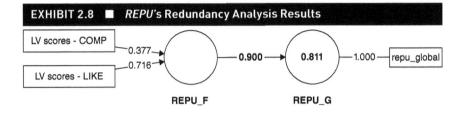

EXHIBIT 2.8 ■ *REPU's* Redundancy Analysis Results

In terms of the structural model relationships, we find that the results of the reflective-formative higher-order construct also meet all structural model evaluation criteria (Hair, Hult. Ringle, & Sarstedt, 2022). All structural model relationships are significant ($p \leq 0.01$). The antecedent constructs *QUAL* (0.440) and *PERF* (0.207) have the strongest effects on *REPU*, while *CSOR* (0.139) and *ATTR* (0.150) are less relevant (Exhibit 2.7). *REPU* itself has a strong effect on *CUSA* (0.537), which, in turn, is strongly related to *CUSL* (0.514). The direct relationship between *REPU* and *CUSL* is somewhat weaker (0.326). The R^2 values of all the endogenous constructs (i.e., 0.700 for *REPU*, 0.288 for *CUSA*, and 0.550 for *CUSL*) are relatively high when taking the number of antecedent constructs into account. The results from PLS$_{predict}$ (Shmueli, Ray, Velasquez Estrada et al. 2016; Shmueli, Sarstedt, Hair, Cheah et al. 2019) support the model's predictive power with regard to its final target construct *CUSL* as all its indicators achieve lower RMSE values than the linear benchmark model.

Finally, we assess which model—with or without the higher-order construct *REPU*—has a better model fit. To do so, go to the **Results report** of the PLS-SEM algorithm and save the model estimates of the **HCM reflective-formative 2nd stage** model by clicking on the **Save** button in the toolbar and choosing a self-explanatory name (e.g., **HCM reflective-formative 2nd stage – PLS**). Next, return to the SmartPLS Workspace, open the **Corporate reputation model**, estimate it using the default settings and also save the results with a self-explanatory name (e.g., **Corporate reputation model – PLS**). Next, in the **Results Report** of the **Corporate reputation model**, click on the **Compare** button in the toolbar. Under **Saved reports → Example corporate reputation (advanced)**, select the saved report **HCM reflective-formative 2nd stage – PLS**. In the upper right area of the **Compare** view that opens, check the box next to **Synchronize navigation**. Then, on the left side, click on one of the **Select detail** button above the displayed model and navigate to **Quality Criteria → Model selection criteria** to obtain the results as displayed in Exhibit 2.9. The BIC value of the target construct **CUSL** is lower for the **Corporate reputation model** (–261.602) than for the **HCM reflective-formative 2nd stage** model (–258.515).

Analogously, we compare the models in terms of their predictive power. To do so, go to **Calculate → PLSpredict** and run the procedure using the default settings. In the **Results report**, go to **Final results → CVPAT → PLS-SEM vs. Indicator average (IA)**. We find that for the model's target construct *CUSL*, the corporate reputation model's PLS-SEM results have a smaller average loss (as an expression of the prediction

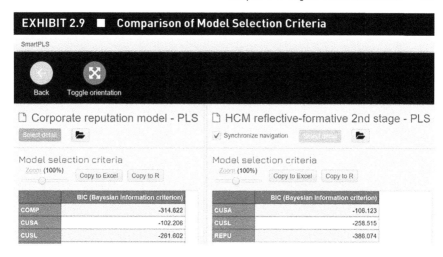

EXHIBIT 2.9 ■ **Comparison of Model Selection Criteria**

error) over the indicator average benchmark—as expressed by an average loss difference of PLS-SEM and the indicator average (IA) with a value of −0.561. The superior predictive capabilities of the PLS-SEM results compared to the IA benchmark also apply to the reflective-formative HCM model and are slightly more pronounced (i.e., with a value of −0.576 for the average loss difference of PLS-SEM and IA). The same finding holds for *CUSL* and the **PLS-SEM vs. Linear model (LM)** outcome with −0.082 for the **Corporate reputation model** and −0.083 for the **HCM reflective-formative** model. To summarize, the results suggest that both model specifications perform very similarly with regard to model fit and predictive power. Hence, researchers may choose either specification, for example, depending on whether they want to include reputation as a distinct concept in their model.

SUMMARY

- **Understand the usefulness of higher-order constructs.** A higher-order construct embraces a more general component (i.e., HOC), measured at a higher level of abstraction, while simultaneously including several subcomponents (i.e., the LOCs), which cover more concrete traits of this conceptual variable under consideration. Higher-order constructs enable reducing the number of structural model relationships, making the PLS path model more parsimonious, while increasing the bandwidth of content covered by the respective constructs.

- **Appreciate the different types of higher-order constructs and understand how to specify them in PLS-SEM.** The use of higher-order constructs builds on carefully established theoretical and conceptual considerations. On these grounds, researchers choose from four major higher-order construct types. Each of these types depicts the specific relationship between the HOC and the LOCs as well as the measurement model used to operationalize the constructs on the lower-order level: reflective-reflective, reflective-formative, formative-reflective, and formative-formative. Generally, the HOC of a reflective-reflective and formative-reflective higher-order construct represents a more general construct that—similar to reflective measurement models—simultaneously explains all the underlying LOCs. Conversely, the HOC is formed by the LOCs in reflective-formative and formative-formative higher-order constructs, which are similar to formative measurement models. To specify a higher-order construct, researchers should draw on the disjoint two-stage approach. When estimating higher-order constricts in PLS-SEM, researchers need to consider further aspects, which relate to the PLS-SEM algorithm weighting scheme, and the use of Mode A and Mode B weighting.

- **Comprehend how to estimate higher-order constructs using the SmartPLS software and how to interpret the results.** Researchers can use SmartPLS to model any of the four higher-order construct types introduced in this chapter. When analyzing the results using the disjoint two-stage approach, researchers need to establish the constructs' reliability and validity in Stage 1. Stage 2 concerns the measurement model validation of the higher-order construct and the structural model assessment.

REVIEW QUESTIONS

1. What is a higher-order construct? Describe each of the four different types of higher-order constructs introduced in this chapter.

2. Which criteria are relevant in the assessment of the different higher-order construct types?

3. What are the consequences of having substantially different numbers of indicators in the LOCs when specifying the higher-order construct using the embedded two-stage approach?

4. Should discriminant validity between lower- and higher-order components be evaluated when using the embedded two-stage approach?

5. Which criteria can researchers draw upon when comparing a model with a higher-order construct to a model without the higher-order construct?

CRITICAL THINKING QUESTIONS

1. Discuss the advantages and disadvantages of higher-order constructs.

2. Can every concept be measured at different levels of abstraction?

3. Screen the literature and identify concepts that are commonly measured using a higher-order construct.

4. When would we use the embedded vs. the disjoint two-stage approach?

5. Do higher-order constructs help resolve discriminant validity problems in PLS-SEM?

KEY TERMS

Bandwidth-fidelity tradeoff

Bottom-up approach

Collect model

Disjoint two-stage approach

Embedded two-stage approach

Extended repeated indicators approach

Formative-formative higher-order
   construct

Formative-reflective higher-order
   construct

Hierarchical common factor model

Hierarchical component model

Higher-order component (HOC)

Higher-order construct

Higher-order model

Jangle fallacy

Lower-order components (LOCs)

Multiple battery model

Reflective-formative higher-order
   construct

Reflective-reflective higher-order
   construct

Repeated indicators approach

Second-order construct

Spread model

Stand-alone higher-order construct

Top-down approach

Type I higher-order construct

Type II higher-order construct

Type III higher-order construct

Type IV higher-order construct

SUGGESTED READINGS

Agarwal, R., & Karahanna, E. (2000). Time flies when you're having fun: Cognitive absorption and beliefs about information technology usage. *MIS Quarterly, 24*(4), 665–694.

Becker, J.-M., Cheah, J. H., Gholamzade, R., Ringle, C. M., & Sarstedt, M. (2023). PLS-SEM's most wanted guidance. *International Journal of Contemporary Hospitality Management, 35*(1), 321–346.

Becker, J.-M., Klein, K., & Wetzels, M. (2012). Hierarchical latent variable models in PLS-SEM: Guidelines for using reflective-formative type models. *Long Range Planning, 45*(5–6), 359–394.

Lohmöller, J.-B. (1989). *Latent variable path modeling with partial least squares. Physica*, Chapter 3.

Sarstedt, M., Hair, J. F., Cheah, J.-H., Becker, J.-M., & Ringle, C. M. (2019). How to specify, estimate, and validate higher-order models. *Australasian Marketing Journal, 27*(3), 197–211.

Sarstedt, M., Hair, J. F., Pick, M., Liengaard, B. D., Radomir, L., & Ringle, C. M. (2022). Progress in partial least squares structural equation modeling use in marketing in the last decade. *Psychology & Marketing, 39*(5), 1035–1064.

Wetzels, M., Odekerken-Schroder, G., & van Oppen, C. (2009). Using PLS path modeling for assessing hierarchical construct models: Guidelines and empirical illustration. *MIS Quarterly, 33*(1), 177–195.

Wold, H. (1982). Soft modeling: The basic design and some extensions. In K. G. Jöreskog & H. Wold (Eds.), *Systems under indirect observations: Part II* (pp. 1–54). North-Holland.

Wilson, B. (2010). Using PLS to investigate interaction effects between higher order branding constructs. In V. Esposito Vinzi, W. W. Chin, J. Henseler, & H. Wang (Eds.), *Handbook of partial least squares: Concepts, methods and applications in marketing and related fields* (pp. 621–652). Springer.

ADVANCED MODELING AND MODEL ASSESSMENT

CHAPTER PREVIEW

Standard conceptualizations that underpin cause-effect relationships in applications of PLS-SEM imply that constructs affect one another in a linear manner. In some instances, however, this assumption does not hold when relationships are nonlinear (e.g., Ahrholdt, Gudergan, & Ringle, 2019). When the relationship between two constructs is nonlinear, the size of the effect between two constructs depends not only on the magnitude of change in the exogenous construct but also on its value. For example, Steiner, Siems, Weber, and Guhl (2014) find a nonlinear relationship between quality satisfaction and customer retention. Specifically, the relationship shows decreasing returns to scale for increasing levels of quality satisfaction (i.e., a degressive shape), while for very high levels of quality satisfaction, returns slightly increase again. Overall, the authors conclude that the relationship between quality satisfaction and customer retention is subject to a saturation effect, implying that customer retention cannot just keep on increasing unlimitedly, even when customer satisfaction with regard to quality continues to increase.

When analyzing nonlinear effects (Wold, 1982), researchers tended to make an assumption of linearity regarding the nature of the effect. While an abundance of different effect types is possible, quadratic effects are most common. This chapter provides an overview of this effect type in PLS-SEM and describes its implementation and assessment using SmartPLS 4.

Another fundamental concern in many PLS-SEM analyses is avoiding measurement model misspecification (e.g., specifying the measurement model as reflective when it should be formative). Incorrect specification of measurement models can result in inaccurate estimates of the parameters. To address the issue of proper measurement model specification, this chapter introduces the confirmatory tetrad analysis for PLS-SEM (CTA-PLS). Drawing on an a priori measurement specification based on theory and logic (see also measurement model operationalization discussed in Chapter 1), the CTA-PLS provides a foundation for empirically assessing whether the data support a formative or a reflective measurement model specification. Research recommends routine use of the CTA-PLS to avoid measurement model misspecification in general (e.g., Bollen & Diamantopoulos, 2017) and in PLS-SEM in particular (e.g., Hair, Sarstedt, Ringle, & Mena, 2012).

NONLINEAR RELATIONSHIPS

In PLS-SEM, relationships between constructs can take various forms. A **linear relationship** can be represented by a straight line (with a positive or negative slope). When a scatterplot between independent and dependent constructs does not indicate a straight relationship, a function with curves can be used to depict the **nonlinear relationship**. Nonlinear relationships occur quite frequently, but determining them a priori on theoretical grounds is not always straightforward. Moreover, linear relationships often approximate nonlinear ones sufficiently well, such that linear relationships appear satisfactory. Yet, if nonlinear relationships are theoretically hypothesized, assessing them empirically when applying PLS-SEM is important.

A typical example of a nonlinear relationship is the diffusion of new products as a function of time (e.g., Bass, 1969). Similarly, the relationship between marketing activities (e.g., advertising spending or promotional activities) and product sales usually follows a nonlinear relationship. Specifically, we would generally expect a positive relationship between marketing activities and sales. However, increasing marketing activities oftentimes involves a positive but diminishing effect on sales (Hay & Morris, 1991). A diminishing effect (nonlinear relationship) implies that after a certain point or threshold every additional unit of marketing activities contributes less to increasing sales. Similarly, another example is the widespread assumption of a positive linear relationship between satisfaction and loyalty, which is likely nonlinear (Eisenbeiss, Cornelißen, Backhaus, & Hoyer, 2014). With higher levels of satisfaction, the strength of the positive effect on loyalty instead decreases at some point. That is, the effect remains positive but is smaller in size.

Establishing nonlinear effects requires careful theoretical reasoning. Also, preliminary analyses can support researchers in identifying nonlinear relationships by plotting values of two variables against each other. In a PLS-SEM analysis, this would involve comparing the scores of two constructs in a scatterplot using spreadsheet software, such as Microsoft Excel, or statistical software, such as IBM SPSS Statistics (see Sarstedt & Mooi, 2019, Chapter 5).

The initial approach to handle nonlinear relationships draws on data transformations of one or both variables between which the relationship is expected or observed. A typical type is **log transformation**, which applies a base 10 logarithm to every observation. As logarithms cannot be calculated for negative values, which by definition occur in the standardized construct scores, researchers need to initially transform the indicators of the corresponding measurement models—rather than the construct scores themselves. If the transformation does not sufficiently linearize the relationship, we have to explicitly include the nonlinear relationship in the PLS path model.

To better grasp the concept of nonlinearity and how to examine nonlinear relationships in PLS path models (Basco, Hair, Ringle, & Sarstedt, 2021; Wold, 1982), consider the following formal (**polynomial**) representation of a **nonlinear effect** as commonly described in a regression context:

$$Y_2 = p_1 Y_1 + p_2 Y_1^2 + \ldots + p_N Y_1^N + z_2.$$

In this equation, Y_2 represents the dependent variable, Y_1 the independent variable, p_1 the path coefficient of the linear relationship between Y_1 and Y_2, and z_2 the dependent variable's error term. The nonlinear effect of the relationship is added by $p_N Y_1^N$ to the linear part $p_1 Y_1$, whereby N stands for the **polynomial degree**, which is the highest exponent occurring in the polynomial. When N equals one, we assume a linear relationship (i.e., $Y_2 = p_1 Y_1 + z_2$), while integer values for N larger than 1 determine different types of nonlinear functions. For example, for $N = 2$, the relationship follows a quadratic form:

$$Y_2 = p_1 Y_1 + p_2 Y_1^2 + z_2.$$

For $N = 3$, the relationship is cubic:

$$Y_2 = p_1 Y_1 + p_2 Y_1^2 + p_3 Y_1^3 + z_2.$$

The polynomial degree determines the form of the nonlinear function, which has approximately $N - 1$ turning points where the slope of the curve changes signs. For example, when $N = 2$, the resulting quadratic function has one turning point (e.g., the slope's negative sign changes to a positive one as shown in the top left box of Exhibit 3.1); when $N = 3$, the resulting cubic function shows two turning points (e.g., with sign changes of the slope from a negative to a positive, and again to a negative one, as shown in the left box of Exhibit 3.2).

Exhibit 3.1 shows alternative forms of a **quadratic effect** with varying values of p_1 and p_2. The p_1 coefficient indicates whether the overall linear relationship between Y_1

EXHIBIT 3.1 ■ **Examples of Quadratic Effects**

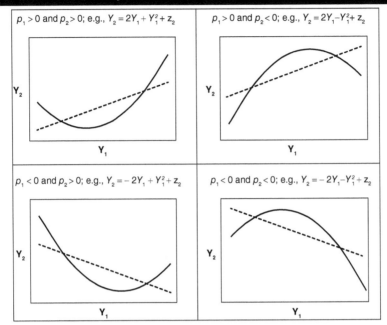

EXHIBIT 3.2 ■ **Examples of Cubic Effects**

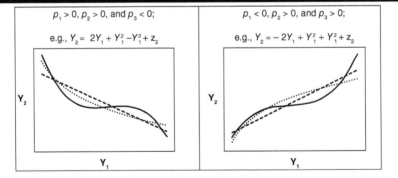

and Y_2 is positive or negative. The nonlinear effect p_2 determines the direction of the curvature. For example, the dotted lines in the two top graphs of Exhibit 3.1 indicate that the overall relationship (i.e., the **linear effect**) between Y_1 and Y_2 is positive. Adding a positive quadratic term creates a convex or u-shaped curve (solid line in the top left box in Exhibit 3.1). Conversely, adding a negative quadratic term creates a concave or an inverse u-shaped curve (solid line in the top right box in Exhibit 3.1). The two bottom graphs in Exhibit 3.1 represent similar situations but with a negative linear effect between Y_1 and Y_2.

Exhibit 3.2 shows a **cubic effect** with varying values of p_1, p_2, and p_3. Compared to the quadratic effect models in Exhibit 3.1, the curves representing the cubic effects have two turning points (i.e., sign changes of their slopes). For the example in the left box of Exhibit 3.2 with positive p_1 and p_2 values but a negative one for p_3, the cubic function is convex at first (left side) but changes into a concave function (right side).

Exhibits 3.1 and 3.2 display typical shapes of nonlinear functions for a wide range of Y_1 and Y_2 values. However, in standard regression analyses, values are frequently limited to a smaller range that includes only certain areas of the overall function. For example, consider a cubic effect as shown in the left box of Exhibit 3.2 and let only the middle part of the function fall into the relevant range of values for the analysis (e.g., approximately from the function's first turning point on the left-hand side to the second turning point on the right-hand side). Then, instead of assuming a cubic function, one would focus only on a slightly increasing function, which almost is linear (with a positive slope). Besides considering the different possible nonlinear functions, the relevant value ranges pose an additional difficulty in making a priori assumptions about the relevant shape of the relationship between two constructs. In PLS-SEM, however, construct scores are usually standardized, which complicates censoring the value range.

In general, hypothesizing quadratic effects already represents a very challenging task. However, theorizing the existence of a cubic effect is even more difficult. For this reason, considering simpler quadratic effects is typically the primary focus when modeling and testing nonlinear effects unless theory provides compelling evidence for more complex nonlinear relationships. In social science models, quadratic effects occur more commonly than other types of nonlinear effects. Therefore, we focus on this type of nonlinear effect in the remainder of this chapter.

Modeling Quadratic Effects in PLS-SEM

To understand how quadratic effects are implemented in PLS path models (e.g., Ahrholdt, Gudergan, & Ringle, 2019), consider the quadratic equation $Y_2 = p_1 Y_1 + p_2 Y_1^2$ in which Y_2 describes the endogenous construct and Y_1 the exogenous construct—note that we omitted the error term for sake of simplicity. This model is specified in a linear way since the path coefficients p_1 and p_2 that link Y_1 and Y_1^2 with Y_2 represent linear relationships (e.g., Sarstedt & Mooi, 2019, Chapter 7). However, by introducing Y_1^2, the relationship becomes nonlinear (i.e., quadratic). The quadratic term comes in the form of an interaction of Y_1 with itself (i.e., $Y_1 \cdot Y_1$) and can therefore be conceived as a special case of a moderation model (Basco, Hair, Ringle, & Sarstedt, 2021; Hair, Hult, Ringle, & Sarstedt, 2022, Chapter 7; Li, 2018) in which Y_1 self-moderates the relationship between Y_1 and Y_2 (Exhibit 3.3). That is, the linear relationship between Y_1 and Y_2 changes in size depending on the values of Y_1.

EXHIBIT 3.3 ■ Nonlinear Effect as Self-Moderation

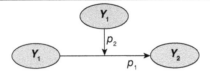

EXHIBIT 3.4 ■ Self-Interaction in PLS-SEM

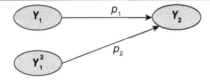

In line with Rigdon, Ringle, and Sarstedt (2010), we can model self-moderation by using an **interaction term,** similar to the interaction effects in standard moderator analysis. Different from a standard moderator analysis, where the interaction term represents the interaction between Y_1 and some other moderator variable, in self-moderation the quadratic term embodies the interaction of Y_1 with itself—as expressed in the following equation: $Y_2 = (p_1 + p_2 Y_1) \cdot Y_1 = p_1 Y_1 + p_2 Y_1^2$. When the quadratic effect is positive, the strength of Y_1's effect on Y_2 increases for higher values of Y_1. Conversely, when the quadratic effect is negative, higher values in Y_1 imply a lower effect of Y_1 on Y_2. Exhibit 3.4 illustrates the concept of self-interaction with a quadratic term (i.e., Y_1^2 with the relationship p_2 to Y_2) as an additional construct in the PLS path model, covering the product of Y_1 with itself, in addition to the linear p_1 relationship from Y_1 to Y_2. Similarly, in the case of a cubic relationship, we would additionally include the construct Y_1^3 with the relationship p_3 to Y_2.

To create the quadratic term Y_1^2 in the PLS path model, the same principles apply as in a standard moderator analysis. Accordingly, researchers can draw on the following approaches (Hair, Hult, Ringle, & Sarstedt, 2022, Chapter 7): the **product indicator approach**, the **orthogonalizing approach**, and the **two-stage approach (nonlinear effects)**. In deciding which approach to apply, it is important to note that the product indicator approach and the orthogonalizing approach are only applicable to reflective measurement models. The indicator multiplication builds on the assumption that the indicators of the exogenous construct and the moderator each stem from a certain construct domain and are in principle interchangeable. Therefore, both approaches are not applicable when the exogenous construct is measured formatively. As an alternative, the two-stage approach is generally applicable, regardless of whether the exogenous construct has a reflective or formative measurement model. Besides its general applicability, simulation studies support its advantageous properties in terms of parameter

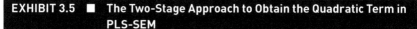

EXHIBIT 3.5 ■ The Two-Stage Approach to Obtain the Quadratic Term in PLS-SEM

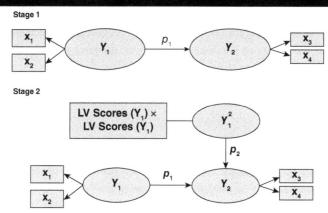

recovery and predictive power (Becker, Ringle, & Sarstedt, 2018; Henseler & Chin, 2010). We therefore recommend using the two-stage approach for generating the interaction term.

The two-stage approach incorporates the following procedure: (1) estimate the model without the interaction term and obtain the construct scores, and (2) use the construct scores as indicators of the independent construct to create the quadratic term in the nonlinear model (LV scores in Exhibit 3.5). Here, the element-wise product of the standardized scores of the independent construct Y_1 with itself serves as an indicator of the quadratic term. The advantage of this approach is that it can be applied to both reflectively and formatively measured independent constructs.

Finally, and similar to the interaction in the moderator analysis (Hair, Hult, Ringle, & Sarstedt, 2022, Chapter 7), it is important to note that the estimated values of p_1 and p_2 represent the strength of the relationship between Y_1 and Y_2 when Y_1 has a value of zero. In many model setups, however, zero is not a number on the scale of Y_1. If this is the case, the interpretation of the quadratic effect becomes problematic. This is the reason why the indicators of the independent variable Y_1 should be mean-centered or standardized. Since PLS-SEM uses standardized data, we focus on this option. Standardization is achieved by subtracting the variable's mean from each observation and dividing the result by the variable's standard error (Sarstedt & Mooi, 2019, Chapter 5). After standardization, Y_1's average value represents the point of reference in the quadratic model, which facilitates interpretation of the effects. Hence, we should standardize the data when estimating nonlinear relationships.

Evaluation of Nonlinear Effects

Measurement model and structural model evaluation criteria also apply to nonlinear models (for more details, see Hair, Hult, Ringle, & Sarstedt, 2022, Chapters 4–6).

The reflectively measured exogenous constructs must meet all relevant criteria in terms of internal consistency reliability, convergent validity, and discriminant validity. Similarly, all formative measurement model assessment criteria must be considered respectively. For the quadratic term, however, there is no such requirement since this construct serves only as an auxiliary measurement, designed to model the quadratic effect. Therefore, similar to the interaction term in standard moderator analysis, the quadratic term's measurement model is not assessed (Hair, Hult, Ringle, & Sarstedt, 2022, Chapter 7). In the two-stage approach, the evaluation criteria apply only to the first stage but not to the second stage since the latter involves only single item constructs.

Following the analysis of the measurement models, the next step requires analyzing the significance of the quadratic effect. The test procedure involves generating the distribution of the parameter by using the bootstrapping procedure (e.g., with 10,000 sub-samples; for more detail, see Hair, Hult, Ringle, & Sarstedt, 2022, Chapter 5). If zero is not included in the 95% confidence interval of the nonlinear effect, it is significantly different from zero at a 5% significance level. In that case, we conclude that the variable of interest has a significant nonlinear effect. Alternatively, we can examine the p value or the corresponding t value to assess whether the nonlinear effect is significant.

When analyzing the significance of effects, we also have to keep in mind that significance does not imply relevance. Particularly with larger sample sizes, very small effects can become significant. However, this does not imply that these effects are relevant. Therefore, the next step is to assess the strength of the nonlinear effect by means of the f^2 effect size. The f^2 effect size assesses the change in the R^2 value of the endogenous construct Y_2 when the exogenous quadratic term Y_1^2 is omitted from the model. The f^2 effect size value is computed as follows:

$$f^2 = \frac{R^2_{\text{model with quadratic effect}} - R^2_{\text{model without quadratic effect}}}{1 - R^2_{\text{model with quadratic effect}}}$$

Consequently, the f^2 effect size indicates how much the quadratic effect contributes to the explanation of the endogenous construct. General guidelines for assessing f^2 suggest that values of 0.02, 0.15, and 0.35 represent small, medium, and large effect sizes, respectively (Cohen, 1988). However, when assessing arguments that apply to moderation effects, Aguinis, Beaty, Boik, and Pierce (2005) point out that the average effect size in tests of moderation is only 0.009. Against that background, Kenny (2015) proposes that 0.005, 0.01, and 0.025 constitute more realistic standards for small, medium, and large effect sizes, respectively. But he also points out that even these values are optimistic given Aguinis, Beaty, Boik, & Pierce's (2005) review. Note again that linear relationships usually offer a reasonable approximation of nonlinear relationships. Additionally, standard considerations concerning statistical power have to be adhered to (see Hair, Hult, Ringle, & Sarstedt, 2022, Chapter 7). Therefore, unless deviations of normality are extreme, researchers should carefully examine whether the inclusion of a quadratic term adds to the analysis of linear effects.

Finally, researchers should also compare the more complex nonlinear model with the more parsimonious linear variant (Basco, Hair, Ringle, & Sarstedt, 2021) using the Bayesian information criterion (BIC). If this analysis suggests the linear model achieves a better balance between model complexity and fit in terms of raw BIC values (Sharma, Sarstedt, Shmueli, Kim, & Thiele, 2019; Sharma, Shmueli, Sarstedt, Danks, & Ray, 2018) or BIC-based Akaike weights (Danks, Sharma, & Sarstedt, 2020), researchers should discard the nonlinear model and interpret the more parsimonious (linear) variant. Alternatively, researchers can compare the model specifications with regard to their predictive power using Liengaard, Sharma, Hult, Jensen, Sarstedt, Hair, & Ringle's (2021) cross-validated predictive ability test (see also Sharma, Liengaard, Hair, Sarstedt, & Ringle, 2023) to empirically justify its consideration.

Results Interpretation

For the interpretation of quadratic effects, consider the following linear model $Y_2 = 0.2 \cdot Y_1$. To better grasp the nature of nonlinear effects, we assume unstandardized effects. Hence, the estimate of p_1 indicates the effect of a one-unit change in Y_1 on Y_2. More precisely, every additional unit of Y_1 increases Y_2 by 0.2 units. Now, if we additionally consider a quadratic effect, the formula changes, for example, into: $Y_2 = 0.6 \cdot Y_1 - 0.15 \cdot Y_1^2$. The coefficient p_1 for the linear Y_1 term changed from 0.2 in the simple linear model to 0.6 in the quadratic model. However, this result does not mean that the linear effect tripled for a unit change in Y_1 since the two estimates of p_1 are not directly comparable. More specifically, the estimate of $p_1 = 0.6$ in the quadratic model does not imply that an increase of one unit of Y_1 changes Y_2 by 0.6, because another Y_1 term exists in the equation (i.e., the quadratic term Y_1^2). Both terms' effects must be considered simultaneously in order to correctly infer the relationship between Y_1 on Y_2. For example, when inserting the values of 0, 1, and 2 for Y_1 into the quadratic model (i.e., $Y_2 = 0.6 \cdot Y_1 - 0.15 \cdot Y_1^2$), we obtain the following results:

$$Y_1 = 0 \Rightarrow Y_2 = 0.6 \cdot 0 - 0.15 \cdot 0 = 0$$
$$Y_1 = 1 \Rightarrow Y_2 = 0.6 \cdot 1 - 0.15 \cdot 1 = 0.45$$
$$Y_1 = 2 \Rightarrow Y_2 = 0.6 \cdot 2 - 0.15 \cdot 4 = 0.60$$

As can be seen, a one-unit change of Y_1 from 0 to 1 entails a change of Y_2 by 0.45 units while a one-unit change from 1 to 2 entails a considerably smaller change of Y_2 by 0.15 units. Hence, in contrast to the linear relationship, the slope and thus the effect of a one-unit change depends on the level of Y_1. For this reason, plotting the quadratic function as shown in Exhibit 3.1 is very useful for presentation and interpretation of the results.

The interpretation of the quadratic effect is similar when considering standardized effects, as is common in PLS-SEM analyses. In this situation, the coefficients indicate the effect of changes in a construct's values with regard to standard deviations, rather than its values as such. More precisely, considering the quadratic equation above, zero represents Y_1's average value as the point of reference for the analysis. If the level of Y_1 is increased (or decreased) by one standard deviation unit, Y_1's linear effect on Y_2 changes by the size of the quadratic term p_2. For example, consider the previous example in which the linear effect p_1 equals 0.60 and the quadratic effect p_2 has a value of -0.15. If the value of the exogenous construct Y_1 increases by one standard deviation unit, one would expect the linear relationship between Y_1 and Y_2 to decrease by 0.15 units. Hence, the effect at the higher level of Y_1 equals $0.60 - 0.15 = 0.45$. These are ceteris paribus considerations, which means that the aforementioned changes are anticipated while everything else in the PLS path model remains constant. Exhibit 3.6 summarizes the rules of thumb for analyzing nonlinear effects in PLS-SEM.

EXHIBIT 3.6   ■   Rules of Thumb for Analyzing Nonlinear Effects in PLS-SEM

- Establish expectations of nonlinear relationships a priori based on theoretical considerations or prior analysis of the relationships between the constructs by means of scatterplots. Keep in mind that linear effects generally offer reasonable approximations of nonlinear effects.

- For the creation of the nonlinear term, use the two-stage approach.

- Create the quadratic term or any other nonlinear effect based on standardized data.

- The exogenous and endogenous variables must be assessed for reliability and validity following the standard evaluation procedures for reflective and formative measures. This does not hold for the nonlinear term, however, which relies on an auxiliary measurement model generated by reusing indicators of the exogenous construct.

- Evaluate whether the nonlinear effects are significant by using the 95% bootstrap confidence intervals. For their significant nonlinear effects, assess their relevance based on the f^2 effects sizes. Outcomes higher than 0.005, 0.01, and 0.025 constitute small, medium, and large f^2 effect sizes respectively.

- Use the BIC to assess whether the more complex nonlinear model achieves a better tradeoff between model complexity and fit compared to the nonlinear model. Alternatively, apply the cross-validated predictive ability test to compare the models with regard to their predictive power.

Case Study Illustration

To illustrate the estimation of nonlinear effects, consider the extended corporate reputation model again. As previously indicated, theory suggests that the effect of customer satisfaction (*CUSA*) on customer loyalty (*CUSL*) is nonlinear (e.g., Eisenbeiss, Cornelißen, Backhaus, & Hoyer, 2014). For this reason, we introduce the quadratic effect of *CUSA* into the model assuming that the positive effect of *CUSA* on *CUSL* diminishes as the level of

CUSA increases. The SmartPLS software offers an option to automatically include a quadratic term. To do so, select the **Quadratic effect** option from the tool bar and select the path relation that should be estimated using a quadratic effect. In our example, left click on the **Click** button that appears on the path relationship between *CUSA* and *CUSL* (Exhibit 3.7, Panel A). SmartPLS will add a yellow circle labeled *QE* to the modeling window, which identifies the quadratic effect and its impact on the link between *CUSA* and *CUSL*. Next, click **Select** in the tool bar and then double-click on quadratic effect *QE* in order to rename it into, for example, *CUSA x CUSA*. Now click on the **Apply** button. To change the shape of this circle so that the label fits well (Exhibit 3.7, Panel B), left-click on the circle. A square with corner points will appear that allows reshaping it.

We can now proceed with the analysis by running the **PLS-SEM algorithm** using the default settings. Note that SmartPLS 4 automatically applies the two-stage approach to generate the interaction term of the quadratic effect. The results in Exhibit 3.8 show that *CUSA* has a direct effect on *CUSL* of 0.467 and that the quadratic term *CUSA x CUSA* has an effect of −0.046, which translates into the following formula:

$$CUSL = 0.467 \cdot CUSA - 0.046 \cdot CUSA \, x \, CUSA$$

To assess whether this quadratic term is significant, run the bootstrapping procedure by going to **Calculate → Bootstrapping** in the tool bar. Run bootstrapping with **10,000** bootstrap samples and default settings in SmartPLS, which include using **Percentile bootstrap** confidence intervals, and a **Two tailed** test at the **0.05** significance level. Next, click on **Start calculation**. In the results report that opens, go to **Final results → Path coefficients → Confidence intervals bias corrected**. We find that the 95% confidence interval of the quadratic effect (*CUSA x CUSA*) is not significant with lower and upper boundaries of −0.093 and 0.000, respectively. This result is also supported by the *p* value, which is 0.054 (**Final results → Path coefficients → Mean, STDEV, T values, p values**). Overall, these results do not provide support for a quadratic effect between *CUSA* and *CUSL*.

Nevertheless, we proceed with the analysis to showcase how to analyze and visualize the quadratic effect in a PLS-SEM analysis. The interpretation of *CUSA*'s quadratic effect uses the direct effect from *CUSA* to *CUSL* with a value of 0.467 as point of reference (Exhibit 3.8). For an increase of *CUSA* by one standard deviation unit, we add the quadratic effect to the linear one. Since the quadratic term is negative, we expect the relationship between *CUSA* and *CUSL* to become weaker (i.e., 0.467 − 0.046) when *CUSA* increases by one standard deviation unit. On the contrary, when *CUSA* decreases by one standard deviation unit, we expect the relationship between *CUSA* and *CUSL* to become stronger (i.e., 0.467 + 0.046). Thereby, we obtain an expected quadratic effect between *CUSA* and *CUSL* similar to the one shown in Exhibit 3.9.

In a further step, we assess whether the quadratic effect is relevant. The small size of the coefficient (i.e., −0.046) in this example already indicates—whether considered

EXHIBIT 3.7 ■ Adding a Quadratic Effect in SmartPLS

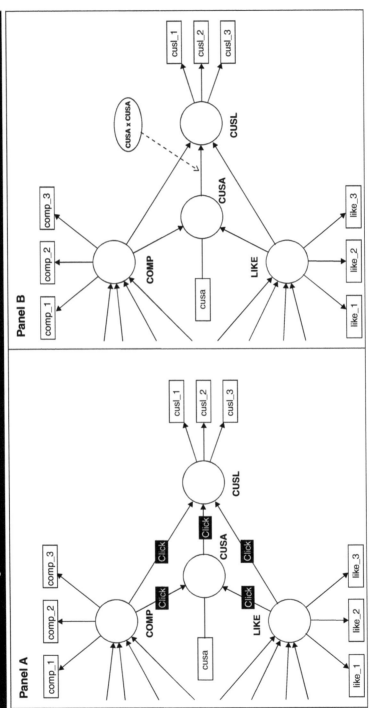

EXHIBIT 3.8 ■ Quadratic Effect Results of *CUSA* in SmartPLS

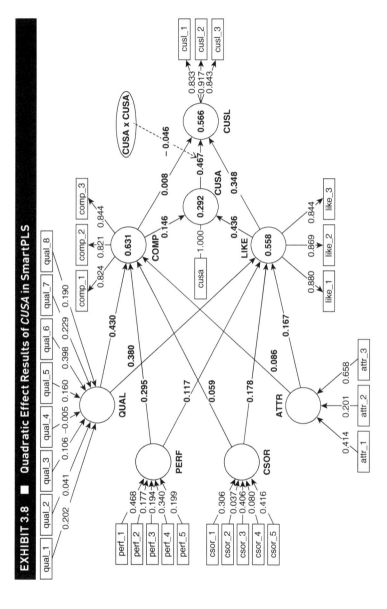

EXHIBIT 3.9 ■ Quadratic Relationship Between Customer Satisfaction and Loyalty

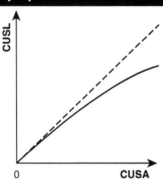

significant or not—that the quadratic term is not particularly relevant. This finding is further supported by the f^2 effect size, which we can request by going to **Quality criteria → f-square** in the standard results report. We find that the quadratic term's f^2 effect size has a value of 0.01. This value falls below the lower limit of 0.02 that, according to Cohen (1988), would at a minimum represent a small effect size. The more liberal interpretation by Kenny (2015), who proposes that values larger than 0.005, 0.01, and 0.025 establish small, medium, and large effect sizes, suggests a medium effect size.

In the final step, we can compare the original model with the extended model that includes the quadratic effect on the grounds of model selection criteria. Specifically, we assess which of the two models produces the lower value in the BIC metric, which has been shown to be particularly effective in PLS-SEM-based model comparison tasks (Sharma, Sarstedt, Shmueli, Kim, & Thiele, 2019). To do so, compute and report the PLS-SEM results of both models and compare their BIC values, which we can find under **Quality criteria → Model selection criteria** in the standard results report. Alternatively, we can make use of SmartPLS's **Compare** option, which allows for directly comparing estimates from two algorithm runs. For this purpose, in the PLS-SEM **Results report** view of the **Nonlinear model**, click on the **Save** button in the toolbar and save this result report with a self-explanatory name (e.g., **Nonlinear model – PLS**). Next, return to the **Workspace** view in SmartPLS, open the **Corporate reputation model**, run the **PLS-SEM algorithm**, and open the **Results report**. Click on the **Save** button in the toolbar and also choose a self- explanatory name (e.g., **Corporate reputation model – PLS**). Next, click on the **Compare** button in the toolbar. Under **Saved reports → Example corporate reputation (advanced)**, select the report **Nonlinear model – PLS**. In the upper right area of the **Compare** view that opens, check the box next to **Synchronize navigation**. Then, click on one of the **Select details** buttons above the displayed models and navigate to **Quality Criteria → Model selection criteria** to obtain the results displayed in Exhibit 3.10. As we can see, the original model produces a lower BIC value for the final target construct *CUSL* (–261.602) than the more complex nonlinear model (–259.299). This result suggests that the more parsimonious original model provides a better fit than

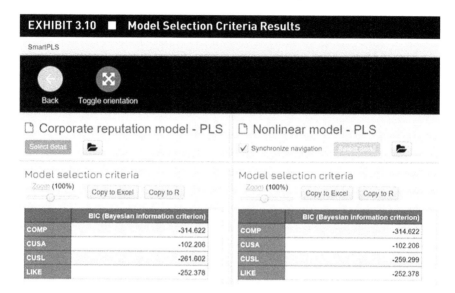

EXHIBIT 3.10 ■ Model Selection Criteria Results

the more complex nonlinear model specification. Finally, when performing the CVPAT comparison of the two alternative models (as described for higher-order constructs in Chapter 2), we find that both models have the same indicator average (IA) loss of –0.561 in the target construct *CUSL*. When considering for the linear model (LM) prediction benchmark, the original model (–0.082) has a higher loss advantage than the nonlinear model alternative (–0.068).

To summarize, we find that *CUSA*'s quadratic effect is nonsignificant and has a relatively low f^2 effect size. While the original model specification produces a better model fit in terms of the BIC, the nonlinear model achieves a higher level of predictive power. In light of these results and considering that the original model is more parsimonious, we would retain the linear specification and discard the nonlinear one.

CONFIRMATORY TETRAD ANALYSIS

Method

Measurement model misspecification is a threat to the validity of SEM results (e.g., Jarvis, MacKenzie, & Podsakoff, 2003). For example, modeling constructs reflectively (i.e., with a **reflective measurement model**) when the conceptualization of the measurement model, and thus the item wordings, follows a formative specification (i.e., a **formative measurement model**) can result in biased outcomes. The reason is that formative indicators are not necessarily highly correlated. In addition, formative indicators produce lower outer loadings when represented in a reflective measurement model. Since indicators with lower outer loadings (< 0.40) should be eliminated from reflectively measured constructs (see Hair, Hult, Ringle, & Sarstedt, 2022, Chapter 4), the incorrect specification of a measurement model as reflective when it should be formative can result in the deletion of indicators that actually should be retained. Any attempt to purify formative measurement models based on correlation patterns among the indicators can have adverse consequences for the constructs' content validity (e.g., Diamantopoulos & Siguaw, 2006). As such, incorrect removal of a formative measurement model's indicator typically has a significant impact on measurement and structural model results.

The primary means to decide whether to specify a measurement model reflectively or formatively is by theoretical reasoning. Guidelines such as those by Jarvis, MacKenzie, and Podsakoff (2003) are helpful in this regard (Exhibit 3.11). Similarly, Bollen and Diamantopoulos (2017, p. 584) provide helpful guidance:

> *Mental experiments are one conceptual tool for determining the nature of an indicator. A researcher should imagine a change in the indicator and ask whether this change is likely to change the value of the latent variable. If so, this is theoretical evidence supporting causal or formative indicators. Alternatively, the researcher should imagine changing the latent variable and ask whether this is likely to change the value of the indicator(s). If so, this favors reflective (effect) indicators. Such mental experiments can provide conceptual support for one type of indicator over the other.*

While the theoretical substantiation of the measurement model specification is useful to avoid measurement model misspecification (Bollen & Diamantopoulos, 2016), there are also empirical means that can help researchers understand whether to specify a measurement model reflectively or formatively. Specifically, Bollen and Ting (1993, 2000) introduced the confirmatory tetrad analysis, which Gudergan, Ringle, Wende, and Will (2008) adapted to PLS-SEM. The **confirmatory tetrad analysis for PLS-SEM (CTA-PLS)** enables researchers to empirically evaluate whether the measurement model specification chosen, based on theoretical grounds, is supported by the data (Rigdon, 2005). More

EXHIBIT 3.11 ■ Decision Rules for Determining Whether a Construct Is Formative or Reflective		
	Formative Model	*Reflective Model*
1. Direction of causality from construct to measure implied by the conceptual definition.	Direction of causality is from the items to the construct.	Direction of causality is from the construct to the items.
Are the indicators (items) (a) defining characteristics or (b) manifestations of the construct?	Indicators are defining characteristics of the construct.	Indicators are manifestations of the construct.
Would changes in the construct cause changes in the indicators?	Changes in the construct do not cause changes in the indicators.	Changes in the construct cause changes in the indicators.

EXHIBIT 3.11 ■ Decision Rules for Determining Whether a Construct Is Formative or Reflective *(Continued)*		
	Formative Model	*Reflective Model*
2. Interchangeability of the indicators/items.	Indicators need not be interchangeable.	Indicators should be interchangeable.
Should the indicators have the same or similar content? Do the indicators share a common theme?	Indicators need not have the same or similar content; indicators need not share a common theme.	Indicators should have the same or similar content; indicators should share a common theme.
Would dropping one of the indicators alter the conceptual domain of the construct?	Dropping an indicator might alter the conceptual domain of the construct.	Dropping an indicator should not alter the conceptual domain of the construct.
3. Covariation among the indicators.	It is not necessary for indicators to covary with each other.	Indicators are expected to covary with each other.
Should a change in one of the indicators be associated with changes in the other indicators?	Not necessarily.	Yes.
4. Nomological net of the construct indicators.	Nomological net of the indicators can differ.	Nomological net of the indicators should not differ.
Are the indicators/items expected to have the same antecedents and consequences?	Indicators are not required to have the same antecedents and consequences.	Indicators are required to have the same antecedents and consequences.

Source: Adapted from Jarvis, MacKenzie, and Podsakoff (2003).

precisely, CTA-PLS can confirm the appropriateness of a reflective measurement model specification. On the other hand, when the method disconfirms the appropriateness of a reflective measurement model, it provides support for a formative measurement model specification. That is, the method enables substantiating the direction of the measurement model relationships or provides support for an alternative specification. Switching the measurement mode of constructs (e.g., from a reflective one to a formative one, and vice versa) solely on the grounds of CTA-PLS results is, however, not meaningful unless additional theoretical or conceptual logic provides support for this change.

The CTA-PLS builds on the concept of **tetrads** (τ), which describe the relationship between pairs of covariances. To better understand what a tetrad is, consider a reflectively measured construct Y_1 with four indicators x_1 to x_4. For this construct, we obtain six covariances (σ) between all possible pairs of the four indicators, as shown in Exhibit 3.12.

EXHIBIT 3.12 ■ Covariances of Four Indicators in a Reflective Measurement Model

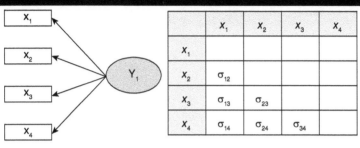

A tetrad is the difference of the product of one pair of covariances and the product of another pair of covariances. The six covariances of four indicator variables result in six unique pairings that form three tetrads:

$$\tau_{1234} = \sigma_{12} \cdot \sigma_{34} - \sigma_{13} \cdot \sigma_{24},$$
$$\tau_{1342} = \sigma_{13} \cdot \sigma_{42} - \sigma_{14} \cdot \sigma_{32}, \text{and}$$
$$\tau_{1423} = \sigma_{14} \cdot \sigma_{23} - \sigma_{12} \cdot \sigma_{43}$$

In reflective measurement models, each tetrad is expected to have a value of zero and, thereby, to vanish—these **vanishing tetrads** are used as input for the analyses. The reason is, according to the domain sampling model (e.g., Nunnally & Bernstein, 1994), reflective indicators represent a single specific concept or trait equally well. Therefore, differences between pairs of covariances of indicators that represent the concept in a similar manner should be zero, provided the domain sampling model holds as assumed by a reflective measurement model. If only one tetrad value in a measurement model is significantly different from zero (i.e., it does not vanish), we should reconsider the reflective measurement model specification and, instead, assume the alternative formative specification—provided that measurement theory supports this step. In other words, the CTA-PLS is a statistical test that considers the hypothesis $H_0: \tau = 0$ (i.e., the tetrad equals zero and vanishes) and the alternative hypothesis $H_1: \tau \neq 0$ (i.e., the tetrad does not equal zero). Therefore, the CTA-PLS initially assumes a reflective measurement specification, while a significant test statistic provides support for H_1. That is, a rejection of the null hypothesis suggests that the measurement model is not reflective and, therefore, likely formative.

Bollen and Ting (2000) provide several numerical examples that illustrate the usefulness of the CTA in the context of covariance-based SEM (CB-SEM). Although the procedures differ, the systematic application of the CTA for assessing measurement models in PLS-SEM is similar to its CB-SEM counterpart (Bollen & Ting, 2000). A necessary requirement for any CTA-PLS is that the items are correlated—at least to some degree. If they were uncorrelated, all tetrads would by definition be zero, which would render the CTA-PLS meaningless. Therefore, a CTA-PLS first requires testing whether at least some of the measurement model's indicators are significantly correlated, for example, by using the IBM SPSS Statistics software (for more details, see Sarstedt & Mooi, 2019, Chapter 5). If this requirement is met, CTA-PLS involves the following main steps:

1. Form and compute all tetrads for the measurement model of a construct.

2. Identify and eliminate redundant tetrads.

3. Perform a statistical significance test whether each tetrad vanishes.

4. Evaluate whether a measurement model's nonredundant tetrads vanish.

In Step 1, all tetrads of the constructs' measurement models are computed. A key consideration is that the tetrad construction requires at least four indicators per measurement model. Otherwise, the CTA-PLS will not produce results for the specific construct. Bollen and Ting (2000) and Gudergan, Ringle, Wende, and Will (2008) provide additional advice regarding how to deal with situations of less than four indicators (i.e., two and three per measurement model), but in PLS-SEM we strongly recommend applying the CTA only on measurement models with at least four indicators. In general, there will be $m!/((m-4)!4!) \cdot 3$ tetrads for measurement models with m indicators (Bollen & Ting, 2000). For example, a measurement model with five, eight, or ten indicators yields 15, 210, or 630 tetrads.

Step 2 of the CTA-PLS identifies and eliminates redundant tetrads. Redundancy exists whenever a tetrad can be represented by two other tetrads. For example, consider the three tetrads extracted from Exhibit 3.12 (i.e., τ_{1234}, τ_{1342}, τ_{1423}). Each tetrad can be represented by the algebraic combination of the other two tetrads (Bollen & Ting, 1993). Thus, one tetrad is redundant and can be eliminated (e.g., τ_{1342}), which leaves two **nonredundant tetrads** for the remaining analysis (e.g., τ_{1234} and τ_{1423}).

While the first two steps of CTA-PLS deal with generating and selecting the (nonredundant) tetrads for each measurement model, Steps 3 and 4 address their significance testing. In Step 3, the CTA-PLS draws on bootstrapping to test whether the nonredundant tetrads differ significantly from zero (see Hair, Hult, Ringle, & Sarstedt 2022, Chapter 5, for further information on the bootstrapping procedure and recommendations for algorithm settings). Analyzing a large number of nonredundant tetrads using bootstrapping requires running a great number of tests. Not only do such complex analyses become increasingly time consuming, but there is an even more severe problem associated with this approach, called **alpha inflation** (also referred to as **multiple testing problem**). This term refers to the fact that the more tests we conduct at a certain significance level, the more likely we are to claim a significant result when this should not be the case (i.e., Type I error). The greater the number of nonredundant tetrads in a particular measurement model, the higher the likelihood that a rejection of the null hypothesis (i.e., a tetrad's value significantly differs from zero) will occur just by chance. For this reason, in Step 4 the CTA-PLS applies a Bonferroni correction that adjusts for alpha inflation. As a result of the Bonferroni correction, the tetrad tests in each measurement model do not assume a significance level of alpha (typically 10%) but alpha divided by the number of nonredundant tetrads. For example, with 10 nonredundant tetrads in a measurement model, each test would assume a significance level of 0.05 / 10 = 0.005. Step 4 calculates the **bias-corrected and Bonferroni-adjusted confidence intervals** of the nonredundant tetrads for a prespecified error level.

A nonredundant tetrad is significantly different from zero if its bias-corrected and Bonferroni-adjusted confidence interval does not include zero. That is, if a measurement model has only vanishing tetrads, we cannot reject the null hypothesis and assume a reflective measurement model specification. On the other hand, if only one of a measurement model's nonredundant tetrads is significantly different from zero (i.e., not all tetrads are vanishing tetrads), one should consider a formative measurement model specification. In any case, it is important to note that when the CTA-PLS does not support a reflective measurement model, adjustments in the model must be consistent with theoretical/conceptual considerations and not solely draw on the empirical test results.

Exhibit 3.13 shows an example of two reflectively measured constructs, Y_1 and Y_2, with four and five indicators, respectively. The measurement model of Y_1 with four indicators requires analyzing two nonredundant tetrads, while the measurement model of Y_2 with five indicators requires considering five nonredundant tetrads.

Exhibit 3.14 shows the (vanishing) tetrads and provides an example of their values as well as the bootstrapping results for 10,000 samples (bootstrap standard error, confidence interval, and p value of every single tetrad). The tetrad values

EXHIBIT 3.13 ■ CTA-PLS Model

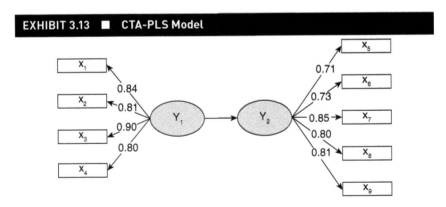

EXHIBIT 3.14 ■ Example CTA-PLS Results				
Y_1	Tetrad value	Bootstrap standard error	$CI_{adj.}$ [a]	p value
τ_{1234}	0.194	0.150	[−0.099; 0.488]	0.194
τ_{1234}	−0.115	0.182	[−0.469; 0.245]	0.527
Y_2	Tetrad value	Bootstrap standard error	$CI_{adj.}$ [a]	p value
τ_{1234}	0.159	0.139	[−0.165; 0.483]	0.254
τ_{1243}	0.223	0.145	[−0.116; 0.558]	0.124
τ_{1235}	0.483	0.142	[0.151; 0.811]	0.001
τ_{1352}	−0.346	0.121	[−0.626; −0.062]	0.004
τ_{1345}	−0.089	0.136	[−0.404; 0.230]	0.512

[a] $CI_{adj.}$ = 90% bias-corrected and Bonferroni-adjusted bootstrap confidence intervals

τ_{1235} and τ_{1352} are significantly different from zero. However, these results do not account for the multiple testing problem. For this reason, the fourth column in Exhibit 3.14 shows the 90% bias-corrected and Bonferroni-adjusted bootstrap confidence intervals. A tetrad may be significantly different from zero when analyzed in isolation but when accounting for the multiple testing problem it may be not significant based on the 90% bias-corrected and Bonferroni-adjusted bootstrap confidence intervals. In the example shown in Exhibit 3.14, we find that the value of two tetrads of Y_2 (i.e., τ_{1235} and τ_{1352}) are significant as zero does not fall into the 90% bias-corrected and Bonferroni-adjusted bootstrap confidence intervals (i.e., these two tetrads do not vanish). The value of tetrad is τ_{1235} significantly higher than zero, while the value of tetrad τ_{1352} is significantly lower than zero. Hence, the CTA-PLS results suggest that Y_2 should be specified as being formative. Note again that this result does not mean that we should mechanically switch to a formative specification, as any change in measurement must be substantiated by theoretical considerations. In contrast, all tetrads of Y_1's measurement model are not significantly different from zero, which empirically substantiates the reflective measurement model specification.

Finally, we would like to express two notes of caution: First and foremost, CTA-PLS is not a silver bullet and interpreting the results requires researchers to carefully consider the specification of the measurement models. Changing the direction of the measurement models solely on the grounds of the CTA-PLS results is unreasonable. Instead, the CTA-PLS results should be interpreted as inputs either to confirm the underlying theoretical and conceptual reasoning or to critically but cautiously reassess the measurement model specification subject to theoretical and conceptual logics. Second, the CTA-PLS results do not allow researchers to draw conclusions about the content validity of the measurement models. That is, the results of CTA-PLS do not indicate whether sufficient parts of a construct domain have been captured by the indicators. This aspect of an empirical analysis requires careful reasoning based on theory and logic, perhaps qualitative research, and potentially with the support of experts who are knowledgeable about the conceptual domain.

The primary rules of thumb on how to conduct CTA-PLS are shown in Exhibit 3.15. In the following case study, we show how to apply the analysis to the corporate reputation model example using the SmartPLS software.

EXHIBIT 3.15 ■ Rules of Thumb for Conducting the Confirmatory Tetrad Analysis

- Establish the mode of measurement (i.e., reflective or formative) on theoretical and conceptual grounds. To do so, also consider qualitative decision rules, such as the four by Jarvis, MacKenzie, and Podsakoff (2003), to support the choice of mode based on theoretical reasoning. Use the CTA-PLS to empirically substantiate your theoretical considerations.

- Consider constructs with at least four indicators in the CTA-PLS, if possible. Otherwise, and only if absolutely necessary, draw on Bollen and Ting (2000) and Gudergan, Ringle, Wende, and Will (2008), who provide some additional advice regarding how to deal with situations of less than four indicators (i.e., two and three per measurement model).

- Inspect the correlations of indicators per measurement model. These correlations must be statistically significant. Otherwise, the tetrad is approximately zero per se and the CTA-PLS becomes meaningless.

- Generate all tetrads per measurement model and exclude the redundant ones. Analyze whether the nonredundant tetrads are significantly different from zero (i.e., whether they vanish).

- Since significance testing accounts for several nonredundant tetrads per measurement at the same time, use the Bonferroni correction to account for the multiple testing problem.

- For significance testing, run the CTA-PLS for a high number of bootstrapping subsamples (e.g., 10,000) and generate the 90% bias-corrected and Bonferroni-adjusted bootstrap confidence interval for each nonredundant tetrad.

- If the confidence interval of a measurement model's tetrad does not include zero, reject the reflective measurement model and assume a formative one. Otherwise, if all of the confidence intervals include zero, the CTA-PLS results empirically provide support for the reflective measurement model specification.

- CTA-PLS results offer guidance regarding the correct measurement model specification, but the final decision of the mode of measurement model always builds on theoretical and conceptual considerations.

Case Study Illustration

Our illustration of the CTA-PLS draws on the corporate reputation model (Hair, Hult, Ringle, & Sarstedt, 2022), which has three constructs with at least four indicators: *CSOR, PERF*, and *QUAL*. To start the analysis, we need to inspect whether the indicators per measurement model have correlations very close to zero. In that case, the tetrad would, per definition, be zero and the analysis becomes meaningless. To obtain the indicator correlations, navigate to the **Example – Corporate reputation (advanced)** project in the **Workspace** and double-click on **Corporate reputation model**. Next, the SmartPLS **Modeling window** opens. Navigate to **Calculate → PLS-SEM algorithm**, which we can find at the top of the SmartPLS screen. Alternatively, we can left-click on the wheel symbol with the label **Calculate** in the tool bar. A combo box opens from which we can select the **PLS-SEM algorithm**. Retain the default settings as described in Hair, Hult, Ringle, and Sarstedt (2022). Before clicking the **Start calculation** button on the lower right side of the dialog box, we need to make sure to tick the box next to **Open report**. Upon completion of the calculations, SmartPLS will open the results report. The left side of the results report contains a list with the different results available. Scroll down this list to **Model and data** and click on **Indicator data (correlations)**. The empirical correlation matrix results open and SmartPLS displays all indicator correlations in the results table on the right side of the **Results report**. We can now inspect whether the indicator correlations of the *CSOR, PERF*, and *QUAL* constructs are sufficiently different from zero. To gain a better overview, we can also copy and paste the correlations into a spreadsheet software, such as Microsoft Excel, by clicking on **Copy to Excel**, for the Excel format, or **Copy to R**, for the R format above the results table. Alternatively, we can click on **Excel** or **HTML** for the respective formats in the tool bar. The results show that the indicators have minimum correlations of 0.406 for *CSOR*, 0.261 for *PERF*, and 0.321 for *QUAL* (Exhibit 3.16). The values are

EXHIBIT 3.16 ■ Indicator Correlations

	csor_1	csor_2	csor_3	csor_4	csor_5
csor_1	1				
csor_2	0.406	1			
csor_3	0.491	0.455	1		
csor_4	0.415	0.473	0.502	1	
csor_5	0.522	0.448	0.540	0.452	1
	perf_1	perf_2	perf_3	perf_4	perf_5
perf_1	1				
perf_2	0.496	1			
perf_3	0.369	0.352	1		

(Continued)

EXHIBIT 3.16 ■ Indicator Correlations *(Continued)*								
	csor_1	*csor_2*	*csor_3*	*csor_4*	*csor_5*			
perf_4	0.395	0.402	0.261	1				
perf_5	0.424	0.383	0.275	0.352	1			
	qual_1	*qual_2*	*qual_3*	*qual_4*	*qual_5*	*qual_6*	*qual_7*	*qual_8*
qual_1	1							
qual_2	0.496	1						
qual_3	0.509	0.530	1					
qual_4	0.461	0.507	0.591	1				
qual_5	0.532	0.399	0.632	0.532	1			
qual_6	0.493	0.425	0.601	0.568	0.604	1		
qual_7	0.424	0.353	0.474	0.502	0.545	0.483	1	
qual_8	0.462	0.321	0.384	0.347	0.387	0.381	0.342	1

clearly different from zero, so we can continue conducting the CTA-PLS. For this purpose, we click on the arrow symbol with the label **Edit** to close the result report and to return to the SmartPLS **Modeling window**.

To run the CTA-PLS, click on **Calculate → Confirmatory tetrad analyses (CTA)** in the SmartPLS **Modeling window**. In the dialog box that opens (Exhibit 3.17), use the default settings that use **10,000 subsamples** for the bootstrapping routine, a two-tailed test (**Two tailed**), and a **Fixed seed**. We generally recommend choosing a **Significance Level** of **0.10** in a CTA-PLS. Check the box next to **Open report** and click on the **Start calculation** button in the lower right of the CTA-PLS dialog box to run the analysis.

Under **Final results** in the list of available CTA-PLS results on the left, we can click on *CSOR, PERF*, and *QUAL*. For each construct, a results table similar to the one shown in Exhibit 3.18 appears. Since we used the fixed seed option, our analysis will produce the same results. Note, however, that when using a random seed, CTA-PLS will produce slightly different results, each time it is run. The reason is that the method builds on bootstrapping, which is a random process that always produces different results when initiated randomly.

The **Original Sample (O)** column shows the results of the nonredundant tetrads. For example, for the construct *CSOR*, whose measurement model has five indicators, the analysis includes five nonredundant tetrads, $\tau_{1234}, \tau_{1243}, \tau_{1235}, \tau_{1352}, \tau_{1345}$. The number of tetrads considered for each construct in the analysis exponentially increases with the

EXHIBIT 3.17 ■ **CTA-PLS Dialog Box**

number of indicators per measurement model. While a measurement model with the minimum number of four indicators includes only two nonredundant tetrads, there are five tetrads for five indicators (*CSOR* and *PERF*) and 20 tetrads for *QUAL*'s measurement model with eight indicators.

The bias-corrected and Bonferroni-adjusted confidence intervals, which SmartPLS uses as default confidence interval types, indicate whether the nonredundant tetrads are significantly different from zero. **CI low adj.** and **CI up adj.** in the results report show the upper and the lower bounds of the 90% bias-corrected and Bonferroni-adjusted confidence intervals. If zero falls into the confidence interval, the tetrad is not significantly different from zero, which implies that it is vanishing. Otherwise, if zero does not fall into the bias-corrected and Bonferroni-adjusted confidence interval, then the nonredundant tetrad is significantly

EXHIBIT 3.18 ■ CTA-PLS Results

CSOR	Original sample (O)	Sample mean (M)	Standard deviation (STDEV)	p value	Bias	CI low	CI up	Alpha adj.	z[1-alpha]	CI low adj.	CI up adj.
1: csor_1,csor_2,csor_3,csor_4	-0.148	-0.146	0.170	0.383	0.003	-0.430	0.128	0.020	2.327	-0.546	0.244
2: csor_1,csor_2,csor_4,csor_3	0.078	0.078	0.135	0.564	0.000	-0.145	0.301	0.020	2.327	-0.237	0.393
4: csor_1,csor_2,csor_3,csor_5	-0.002	-0.001	0.158	0.988	0.003	-0.266	0.254	0.020	2.327	-0.374	0.362
6: csor_1,csor_3,csor_5,csor_2	-0.101	-0.101	0.199	0.613	0.000	-0.427	0.227	0.020	2.327	-0.563	0.362
10: csor_1,csor_3,csor_4,csor_5	-0.012	-0.014	0.145	0.932	-0.002	-0.250	0.228	0.020	2.327	-0.348	0.327
PERF	**Original sample (O)**	**Sample mean (M)**	**Standard deviation (STDEV)**	**p value**	**Bias**	**CI low**	**CI up**	**Alpha adj.**	**z[1-alpha]**	**CI low adj.**	**CI up adj.**
1: perf_1,perf_2,perf_3,perf_4	-0.073	-0.072	0.107	0.495	0.001	-0.251	0.102	0.020	2.327	-0.325	0.175
2: perf_1,perf_2,perf_4,perf_3	-0.038	-0.036	0.096	0.693	0.002	-0.197	0.118	0.020	2.327	-0.262	0.183
4: perf_1,perf_2,perf_3,perf_5	-0.019	-0.018	0.112	0.869	0.000	-0.204	0.166	0.020	2.327	-0.281	0.243
6: perf_1,perf_3,perf_5,perf_2	-0.032	-0.031	0.099	0.742	0.001	-0.196	0.129	0.020	2.327	-0.263	0.196
10: perf_1,perf_3,perf_4,perf_5	0.091	0.089	0.125	0.468	-0.001	-0.114	0.297	0.020	2.327	-0.199	0.383

EXHIBIT 3.18 ■ CTA-PLS Results (Continued)

CSOR / QUAL	Original sample (O)	Sample mean (M)	Standard deviation (STDEV)	p value	Bias	CI low	CI up	Alpha adj.	z[1−alpha]	CI low adj.	CI up adj.
1: qual_1,qual_2,qual_3,qual_4	0.159	0.158	0.138	0.248	−0.001	−0.067	0.386	0.005	2.808	−0.227	0.546
2: qual_1,qual_2,qual_4,qual_3	0.223	0.221	0.141	0.113	−0.002	−0.007	0.456	0.005	2.808	−0.170	0.619
4: qual_1,qual_2,qual_3,qual_5	0.483	0.479	0.142	0.001	−0.004	0.253	0.721	0.005	2.808	0.088	0.887
6: qual_1,qual_3,qual_5,qual_2	−0.346	−0.341	0.122	0.005	0.005	−0.552	−0.150	0.005	2.808	−0.694	−0.008
7: qual_1,qual_2,qual_3,qual_6	0.384	0.380	0.145	0.008	−0.004	0.149	0.626	0.005	2.808	−0.020	0.795
10: qual_1,qual_2,qual_3,qual_7	0.266	0.263	0.138	0.054	−0.002	0.041	0.495	0.005	2.808	−0.119	0.655
13: qual_1,qual_2,qual_3,qual_8	0.116	0.113	0.110	0.292	−0.003	−0.062	0.301	0.005	2.808	−0.190	0.429
17: qual_1,qual_2,qual_5,qual_4	−0.027	−0.024	0.179	0.880	0.003	−0.325	0.265	0.005	2.808	−0.534	0.473
23: qual_1,qual_2,qual_7,qual_4	0.163	0.162	0.171	0.340	−0.001	−0.118	0.446	0.005	2.808	−0.317	0.646
26: qual_1,qual_2,qual_8,qual_4	−0.274	−0.269	0.135	0.043	0.005	−0.501	−0.056	0.005	2.808	−0.658	0.101
30: qual_1,qual_5,qual_6,qual_2	0.137	0.134	0.133	0.305	−0.002	−0.080	0.359	0.005	2.808	−0.235	0.514
33: qual_1,qual_5,qual_7,qual_2	0.087	0.084	0.125	0.489	−0.003	−0.116	0.295	0.005		−0.261	0.440

(Continued)

EXHIBIT 3.18 ■ CTA-PLS Results (Continued)

CSOR	Original sample (O)	Sample mean (M)	Standard deviation (STDEV)	p value	Bias	CI low	CI up	Alpha adj.	z(1-alpha)	CI low adj.	CI up adj.
42: qual_1,qual_6,qual_8,qual_2	-0.172	-0.168	0.136	0.205	0.005	-0.401	0.047	0.005	2.808	-0.560	0.205
73: qual_1,qual_3,qual_7,qual_8	0.050	0.049	0.138	0.720	-0.001	-0.177	0.278	0.005	2.808	-0.338	0.439
85: qual_1,qual_4,qual_6,qual_7	-0.123	-0.123	0.1549	0.411	-0.001	-0.367	0.123	0.005	2.808	-0.540	0.297
97: qual_1,qual_5,qual_6,qual_8	0.054	0.051	0.128	0.675	-0.003	-0.154	0.266	0.005	2.808	-0.302	0.415
100: qual_1,qual_5,qual_7,qual_8	0.077	0.075	0.124	0.534	-0.003	-0.125	0.285	0.005	2.808	-0.269	0.429
110: qual_2,qual_3,qual_6,qual_4	0.248	0.245	0.147	0.091	-0.003	0.010	0.493	0.005	2.808	-0.160	0.663
121: qual_2,qual_3,qual_5,qual_7	0.484	0.479	0.161	0.003	-0.005	0.224	0.755	0.005	2.808	0.037	0.942
156: qual_2,qual_6,qual_7,qual_5	0.095	0.096	0.166	0.567	0.001	-0.179	0.368	0.005	2.808	-0.372	0.561

different from zero, and it is not vanishing. The latter situation occurs when the signs of **CI low adj.** and **CI up adj.** both are negative or both are positive. This is the case for tetrads 4, 6, and 121 of the *QUAL* construct. The numbers stem from the full list of tetrad generation (e.g., 210 tetrads for the measurement model of *QUAL* with eight indicators). This list becomes considerably reduced after determining the nonredundant tetrads in Step 2 of the CTA-PLS. SmartPLS, however, keeps the previous numbers to indicate which (nonredundant) tetrads have been selected for the final analysis. The confidence interval of *QUAL's* tetrad 4 has a lower boundary of 0.088 and an upper boundary of 0.887 and therefore does not include zero. Similarly, with a lower boundary of −0.694 and an upper boundary of −0.008, tetrad 6 has a significant negative value. Finally, the confidence interval of tetrad 121 also does not include zero (i.e., it has a significant positive value). These results suggest that *QUAL's* measurement model is indeed formative, providing support for its original specification.

On the contrary, all nonredundant tetrads of *CSOR* and *PERF* vanish since all the tetrads' confidence intervals include zero. Hence, we assume that these constructs' measurement models are reflective, which opposes the theoretical and conceptual assumptions. In light of these results, we next examine the underpinnings of the measurement model specifications applying Jarvis, MacKenzie, and Podsakoff's (2003) qualitative decision rules (Exhibit 3.11). Our examination of the construct (Exhibit 3.19) suggests these indicators are independent sources that form the construct rather than reflections of a singular concept. In other words, a change in the construct does not necessarily imply a simultaneous change of all indicators. We therefore retain the original formative specification, also in light of the fact that various studies have used this measurement specification in prior research (e.g., Eberl, 2010; Schwaiger, Sarstedt, & Taylor, 2010; Schwaiger, Witmaier, Morath, & Hufnagel, 2021). Remember that the CTA-PLS results should not be mechanically applied, especially since formative indicators can exhibit strong correlations, which should, however, not be so large that they cause collinearity issues (Har, Hult, Ringle, & Sarstedt, 2022, Chapter 5).

To summarize, the CTA-PLS results indicate that *QUAL's* measurement model should be specified as being formative whereas the measurement models of *CSOR* and *PERF* should be specified as being reflective. While these empirical findings match the theoretical and conceptual considerations with regard to *QUAL*, they oppose the original specification for the *CSOR* and *PERF* constructs. However, as the decision on how to specify measurement models primarily relies on theoretical and conceptual considerations, the CTA-PLS results should be taken as cause to critically reexamine, but not necessarily determine, these constructs' measurement model specifications. A reexamination of the conceptual foundations for these two constructs—employing the qualitative decision rules of Jarvis, MacKenzie, and Podsakoff (2003)—however, supports the initial formative measurement model specification for *CSOR* and *PERF*.

EXHIBIT 3.19 ■ Indicators of the Constructs Included in the CTA-PLS

Corporate Social Responsibility (CSOR)	
csor_1	[The company] behaves in a socially conscious way.
csor_2	[The company] is forthright in giving information to the public.
csor_3	[The company] has a fair attitude toward competitors.
csor_4	[The company] is concerned about the preservation of the environment.
csor_5	[The company] is not only concerned about profits.
Performance (PERF)	
perf_1	[The company] is a very well-managed company.
perf_2	[The company] is an economically stable company.
perf_3	The business risk for [the company] is modest compared to its competitors.
perf_4	[The company] has growth potential.
perf_5	[The company] has a clear vision about the future of the company.
Quality (QUAL)	
qual_1	The products/services offered by [the company] are of high quality.
qual_2	[The company] is an innovator, rather than an imitator with respect to [industry].
qual_3	[The company]'s products/services offer good value for money.
qual_4	The services [the company] offered are good.
qual_5	Customer concerns are held in high regard at [the company].
qual_6	[The company] is a reliable partner for customers.
qual_7	[The company] is a trustworthy company.
qual_8	I have a lot of respect for [the company].

SUMMARY

- **Comprehend the basic concepts of nonlinear analysis in a PLS-SEM context.**
  Nonlinear effects occur when the relationship between two constructs does
  not follow a straight line but a curve when plotting the constructs' values in a
  scatterplot. When the relationship between two constructs is nonlinear, the size of
  the effect between two constructs is not constant and depends on the values and
  the magnitude of change in the exogenous construct. When analyzing nonlinear
  effects, researchers have to make an assumption regarding the nature of the effect.

While numerous different effect types are possible, quadratic effects are most common.

- **Learn how to use the SmartPLS software to estimate quadratic effects.** Analyzing a quadratic effect in PLS-SEM requires researchers to include an interaction term that accounts for the self-moderation of the exogenous construct. To implement the interaction term in a PLS path model, the two-stage approach should be used. To assess the quadratic effect, researchers need to evaluate the interaction term's effect regarding its significance and relevance (e.g., using the f^2 effect size) and compare the BIC of the linear model with the one of the nonlinear model. If the results support a nonlinear model, the researcher should analyze the resulting function in its interpretable range and decide whether the positive/negative and increasing/decreasing or (inverse) u-shaped relationship meets the a priori expectations.

- **Assess the measurement model mode with the confirmatory tetrad analysis in PLS-SEM (CTA-PLS).** CTA-PLS is an approach to empirically evaluate a construct's measurement model mode (i.e., formative or reflective). The method requires, ideally, at least four indicators per measurement model. In case of a reflective measurement model, all its nonredundant tetrads are zero and, therefore, vanish. However, if at least one of the nonredundant tetrads is significantly different from zero, we should consider rejecting the reflective measurement model and, instead, assume a formative specification. When evaluating the measurement model mode, researchers should primarily rely on theoretical and conceptual considerations along with the empirical evidence provided by the CTA-PLS.

- **Understand how to run the CTA-PLS in SmartPLS and how to interpret the results.** A CTA-PLS—as executed in SmartPLS—typically draws on a large number of bootstrap samples (typically 10,000) and assumes a 10% significance level. Based on the bootstrapping results, SmartPLS returns the bias-corrected and Bonferroni-adjusted confidence intervals for the nonredundant tetrads of all measurement models with four or more indicators. The results facilitate analyzing whether a nonredundant tetrad is statistically significant (i.e., whether it vanishes).

REVIEW QUESTIONS

1. Explain the concept of self-interaction in the context of modeling quadratic effects.

2. What is the most versatile approach for creating a quadratic term with regard to the measurement model of the exogenous construct?

3. How do you assess the results of a nonlinear model?

4. What is a tetrad, and how would you compute it?

5. What are the key steps of CTA-PLS?

6. How do you decide whether a measurement model is reflective or formative based on the CTA-PLS results?

CRITICAL THINKING QUESTIONS

1. Give an example of a quadratic effect.

2. Explain what the path coefficients in a nonlinear model mean.

3. Discuss the steps when analyzing a nonlinear effect in PLS-SEM.

4. Why is measurement model misspecification a critical problem?

5. Use Jarvis, MacKenzie, and Podsakoff's (2003) qualitative decision criteria to derive sets of reflective and formative indicators for measuring customer satisfaction with your mobile phone service provider.

6. If the CTA-PLS results do not match the assumed measurement model specification, how do you finally decide which kind of measurement mode to use?

KEY TERMS

Alpha inflation

Bias-corrected and Bonferroni-adjusted confidence intervals

Confirmatory tetrad analysis for PLS-SEM (CTA-PLS)

Cubic effect

Formative measurement model

Interaction term

Linear effect

Linear relationship

Log transformation

Measurement model misspecification

Multiple testing problem

Nonlinear effect

Nonlinear relationship

Nonredundant tetrads

Orthogonalizing approach

Polynomial

Polynomial degree

Product indicator approach

Quadratic effect

Reflective measurement model

Tetrads (τ)

Two-stage approach (nonlinear effects)

Vanishing tetrads

SUGGESTED READINGS

Ahrholdt, D. C., Gudergan, S., & Ringle, C. M. (2019). Enhancing loyalty: When improving consumer satisfaction and delight matters. *Journal of Business Research*, *94*(1), 18–27.

Basco, R., Hair, J. F., Ringle, C. M., & Sarstedt, M. (2021). Advancing family business research through modeling nonlinear relationships: Comparing PLS-SEM and multiple regression. *Journal of Family Business Strategy*, *13*(3), 100457.

Bollen, K. A., & Ting, K.-F. (2000). A tetrad test for causal indicators. *Psychological Methods*, *5*(1), 3–22.

Gudergan, S. P., Ringle, C. M., Wende, S., & Will, A. (2008). Confirmatory tetrad analysis in PLS path modeling. *Journal of Business Research*, *61*(12), 1238–1249.

Rigdon, E. E., Ringle, C. M., & Sarstedt, M. (2010). Structural modeling of heterogeneous data with partial least squares. In N. K. Malhotra (Ed.), *Review of marketing research* (pp. 255–296). M. E. Sharpe.

Sarstedt, M., Hair, J. F., Pick, M., Liengaard, B. D., Radomir, L., & Ringle, C. M. (2022). Progress in partial least squares structural equation modeling use in marketing in the last decade. *Psychology & Marketing*, *39*(5), 1035–1064.

4 ADVANCED RESULTS ILLUSTRATION

LEARNING OUTCOMES

1. Understand why and how to run a necessary condition analysis (NCA) in PLS-SEM.

2. Comprehend how to use SmartPLS to run an NCA in PLS-SEM.

3. Develop and interpret the importance-performance map analysis (IPMA).

4. Comprehend how to conduct the IPMA using the SmartPLS software.

CHAPTER PREVIEW

When researchers apply PLS-SEM to analyze models with causal-predictive relationships, they explicitly or implicitly refer to an additive sufficiency logic. According to this logic, each of the antecedent constructs in a structural model is sufficient (but not necessary) for producing changes in the dependent construct. The reasoning is that values in single constructs can compensate for values (or lack thereof) in other constructs. For instance, in a PLS-SEM study, Carlson, Gudergan, Gelhard, and Rahman (2019) unpack how customer engagement is linked to the customers' intentions to share information with the brand. They suggest that each of four customer engagement dimensions (i.e., absorption, enthusiasm, focused attention, and interaction) stimulates intentions to share with the brand. Similarly, in PLS-SEM studies on the sources of competitive advantages, each antecedent is assumed to contribute to higher performance and if one source is absent, another source can compensate for it–see, for example, Richter, Schlaegel, Midgley, and Tressin's (2019) PLS-SEM study on global sourcing success.

However, this logic might not always resemble the theorizing that authors aim to test empirically. Consider an omnichannel retailer that operates traditional and online channels. For this retailer both physical store infrastructure and digital infrastructure are necessary to operate and to generate revenue. If either of these were not there or were severely underdeveloped, the omnichannel retailer could not operate and would generate insufficient revenue. In contrast, in online-only retail, physical store infrastructure is not necessary. Hence, we have to consider must-have and should-have conditions to understand what drives retailer performance. The existence of both (i.e., of necessary must-have conditions and sufficient should-have conditions) is common in many fields of research. This additional necessity logic can be tested by using the **necessary condition analysis** (**NCA**; Dul, 2016, 2020). Although the NCA is a relatively new methodology, it has already gained considerable attention in management research as a means or identifying necessary conditions (e.g., Bergh, Boyd, Byron, Gove, et al., 2022; Bokrantz & Dul, 2023; Fainshmidt, Witt, Aguilera, & Verbeke, 2020; Hauff, Guerci, Dul, & van Rhee, 2019). In PLS-SEM, NCA results can also support researchers and practitioners in their decision making. Especially if theoretically established relationships in the structural model are not statistically significant, the NCA can add information about these relationships' necessity. This chapter details the meaning of sufficiency logic and necessity logic, the fundamentals of an NCA, and how to use the method in a PLS-SEM context. We conclude with a practical NCA application drawing on our corporate reputation example and the SmartPLS software.

The second topic we cover in this chapter is the importance-performance map analysis (IPMA), which extends the standard results reporting of path coefficient estimates, R^2, and similar parameters by adding a visual interpretation based on the average values of the construct scores. More precisely, the IPMA compares the total effects, representing a predecessor constructs' total influence on a specific target construct, with their average construct scores. The analysis enables identification of predecessor constructs that have a relatively high (low) importance in explaining the target construct (i.e., those that have a strong [weak] total effect), but also have a relatively low (high) performance (i.e., low [high] average construct scores). These findings are particularly useful for interpreting where to emphasize efforts to improve the performance of a key target construct predicted in the PLS path model.

NECESSARY CONDITIONS ANALYSIS

Purpose and Method

The aim of any PLS-SEM is to estimate and empirically substantiate theoretically established causal-predictive relationships (Sarstedt, Ringle, & Hair, 2021; Wold 1982). PLS-SEM results reveal how certain conditions (X) lead to a given outcome (Y). Researchers who interpret their PLS-SEM findings usually employ expressions such as "X increases Y" or "a higher X leads to a higher Y" (e.g., Richter, Schlaegel, Midgley, & Tressin., 2019). The interpretation of relationships between the conditions and the outcome therefore follows a **sufficiency logic** (Dul, 2016). Understanding relationships in terms of a sufficiency logic is highly relevant in testing theory and in deriving practical implications from the data analysis. Such an analysis in marketing research would, for example, aim to identify the conditions that lead to a stronger intention to repurchase from an online retailer. Based on the proposition that greater enjoyment leads to greater online spending, we could, for instance, assume that in online buying customer enjoyment is a should-have condition. That is, following a sufficiency logic, enjoyment may be sufficient to trigger online spending. However, this condition might not be necessary because the absence of enjoyment could be compensated for by other conditions such as product availability or pricing.

In contrast, a **necessity logic** implies that an outcome—or a certain level of an outcome—can only be achieved if the necessary condition (e.g., a certain indicator or construct) is in place or has achieved a certain level. If the necessary condition is not in place, the outcome will not materialize (i.e., the condition is necessary but not sufficient for an outcome). Consequently, the necessary condition must be satisfied to achieve a certain outcome, while other conditions cannot compensate for the absence of a necessary condition. For example, product availability may be a necessary but not always sufficient condition for online shopping. To express such a necessity, researchers rely on expressions such as "X is needed for Y," "X is a precondition for Y," or "Y requires X" (Dul, 2021, Chapter 2). Accordingly, the necessary condition, which can represent a constraint, a bottleneck, or a critical condition, must be satisfied to achieve a certain outcome Y or a specific level of Y (i.e., Y will not be achieved if X or a certain level of X is not in place). Referring to our previous example, a sufficiency perspective would aim to identify the degree of customer enjoyment necessary to achieve a certain level of online spending.

The necessity logic complements the should-have conditions, by disclosing potential must-have conditions that need to be met to produce a certain outcome. Considering both sufficiency and necessity perspectives is highly relevant for business research across a range of disciplines (e.g., Dul, Hak, Goertz, & Voss, 2010). For instance, researchers and practitioners often seek to understand the conditions that lead to customers repurchasing, as well as the conditions that are necessary for this behavior. Therefore, they aim to determine the conditions that produce the best possible outcome (i.e., the should-have conditions; sufficiency logic), as well as those that are critical for a particular outcome (i.e., the must-have conditions; necessity logic). Ignoring or neglecting necessary conditions, which we theoretically assume exist, yields incomplete, and possibly misleading, findings and recommendations.

To analyze a model from a necessity perspective, researchers can rely on the NCA (Dul, 2016, 2020), which contrasts variable scores of dependent and independent variables to identify necessary condition thresholds that must be met to obtain a certain outcome. Correspondingly, the implementation of the NCA in a PLS-SEM context uses the construct scores produced by the standard algorithm as input to explore necessary conditions in the structural model relationships. In using these scores, we focus on bivariate relationships in that a necessity condition between a dependent and independent construct does not depend on other conditions in the same estimation.

In the following section, we discuss the concept of an NCA by first highlighting how to identify the necessary conditions and quantify their size. Next, we discuss the results interpretation, after which we mention important matters to consider in running an NCA. Exhibit 4.1 summarizes our recommendations. Moreover, Richter, Schubring, Hauff, Ringle et al. (2020), and Richter et al. (2023) explain in further detail how to use the NCA in PLS-SEM (see also Duarte, Silva, Linardi, & Novais, 2022; Richter, Hauff, Kolev, & Schubring, 2023; Sukhov, Olsson, & Friman, 2022).

EXHIBIT 4.1 ■ Rules of Thumb for Conducting an NCA

- Preferably use unstandardized construct scores obtained from PLS-SEM to run the NCA.

- Assume that there is a necessity condition when the necessity effect size d is 0.1 or larger. Apply the permutation test to assess whether d is significant.

- Analyze the ceiling regression – free disposal hull (CR-FDH) ceiling line charts and the bottleneck tables to identify values in the independent constructs needed to produce certain outcome levels.

Identifying Necessary Conditions

To disclose the necessary conditions, the NCA relies on scatterplots of the dependent and independent constructs' scores. While the standard PLS-SEM analysis establishes a linear function through the center of all data points, the NCA maps the ceiling line of the data. Exhibit 4.2 shows a scatterplot along with ceiling lines for the (hypothetical) relationship between a dependent construct Y and an independent construct X. The ceiling lines separate the potential area with observations (S; also called the *scope*), as represented by the entire square in Exhibit 4.2 into an area with observations (the lower right area in Exhibit 4.2) and an area without observations (C; also called the *ceiling zone*), as represented by the upper left in Exhibit 4.2. Most data points in the scatterplot occupy the lower right side of the figure, while the upper left side includes only a few, which implies that X puts a constraint on Y for these combinations of construct scores. For example, Y cannot achieve high values if there are low values of X.

Mostly, there are two default ceiling lines. One is the **ceiling envelopment – free disposal hull (CE-FDH) line**, which is a step function through the upper left data points as shown in Exhibit 4.2. The other is the **ceiling regression – free disposal hull (CR-FDH) line**, which is a simple linear regression line through the data points of the CE-FDH line (Exhibit 4.2). The CE-FDH ceiling line is recommended for discrete data or if the pattern of observations near the ceiling line has many steps of different magnitudes. The CR-FDH line is useful for continuous data and if the pattern of observations near the ceiling line is approximately linear—as in Exhibit 4.2. Since we use continuous data in PLS-SEM, we generally focus on the CR-FDH ceiling line.

The ceiling line specifies the minimum level of X that is necessary to achieve a certain level of Y. To grasp this concept, Exhibit 4.3 shows an example of an NCA ceiling line chart from SmartPLS 4. In this example, for values of X below 3.0, the CR-FDH ceiling line identifies no observations in which Y takes a value of 4.44 or more. Thus, the independent variable X must have a value of at least 3.0 for Y to take a value of 4.44. This NCA finding differs from the interpretation of linear regression, resembling PLS-SEM, where an increase in the independent variable leads to a certain average increase of the dependent variable. The size of the ceiling zone C in Exhibit 4.3 depends on the observations that determine the ceiling line. However, single (outlier) observations, which would appear in the left upper corner of Exhibit 4.3, can have a strong impact

EXHIBIT 4.2 ■ Ceiling Lines Example

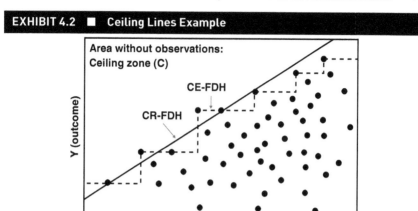

EXHIBIT 4.3 ■ Example of an NCA ceiling line chart in SmartPLS

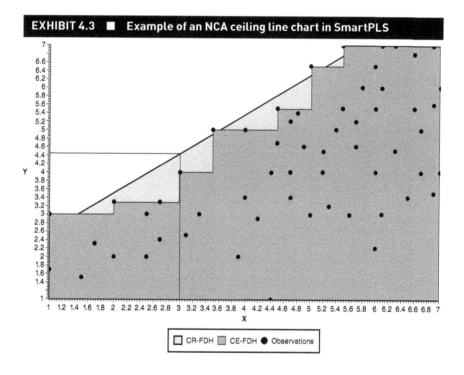

on the outcome. For this reason, Dul's (2021, Chapter 3) recommendations for outliers treatment in an NCA should receive special attention.

Another way of illustrating the NCA results is by means of a **bottleneck table**. In such a table, the first column shows the dependent variable or the outcome, whereas the next (and additional) column(s) represent(s) the condition(s) that must be satisfied to achieve the outcome. The results of both the outcome and the condition(s) can refer to the actual values, percentage values of the range, or percentiles. When applying an NCA in a PLS-SEM context, the actual values (and their transformations, such as percentage ranges or percentiles) can take different forms, depending on the construct scores used in the analysis.

Exhibit 4.4 illustrates a bottleneck table referring to the above example (see Exhibit 4.3). For instance, Exhibit 4.4 shows that for Y to achieve a level of 4.0 (second column), the independent variable X must achieve a level of 2.5 (third column). Additionally, NN indicates that the independent variable is not necessary for this level of the dependent construct. For example, the condition X is not necessary to accomplish an outcome level of 2.2 (or lower) in Y. However, for higher outcome levels of Y, the condition X becomes necessary (Exhibit 4.4). The first column lists the percentage ranges for the outcome, which is a default visualization often used in an NCA. The column includes the values of Y in percentages of their ranges, where 0 corresponds to the lowest *observed* value, and 100 to the highest *observed* value. For instance, to achieve an outcome level of 70% (first column), which is indicated by an actual value of 5.2 on our 7-point scale (second column), X has to be at a level of 3.8 (third column).

In addition, Exhibit 4.4 presents X in counts (column 4) and in percentiles (column 5). Displaying X in the CR-FDH bottleneck table in terms of counts focuses on the number of observations in the dataset that do not meet the necessary level of X to accomplish a certain level of Y. For instance, if we consider an outcome level of 5.2 for

EXHIBIT 4.4 ■ Bottleneck Table (CR-FDH)				
Y in Percentiles	*Y in Actual Values*	*X in Actual Values*	*X in Counts*	*X in Percentiles*
0.0	1.0	NN	0	0.0
10.0	1.6	NN	0	0.0
20.0	2.2	NN	0	0.0
30.0	2.8	1.2	2	3.3
40.0	3.4	1.9	4	6.7
50.0	4.0	2.5	8	13.3
60.0	4.6	3.2	12	20.0
70.0	5.2	3.8	14	23.3
80.0	5.8	4.5	20	33.3
90.0	6.4	5.1	29	48.3
100.0	7.0	5.8	38	63.3

Note: NN = a result is not available and, thus, does not represent a necessary condition.

the dependent variable Y, we find that 14 cases do not achieve the necessary level of X (i.e., a level for X of at least 3.8). Similarly, the percentile option displays the percentage of cases that do not meet the necessary level of X to accomplish a certain level of Y. We see, for example, that the 14 cases of X that do not achieve a level of 3.8 represent 23.3% of all 60 cases in this example.

To identify whether a construct is necessary for achieving a certain outcome, researchers should draw on the **necessity effect size d** and its significance. The necessity effect size is computed by dividing the ceiling zone C by the scope S and therefore ranges between $0 \leq d \leq 1$. Dul (2016) suggests that:

- $0 < d < 0.1$ can be characterized as a small effect,

- $0.1 \leq d < 0.3$ as a medium effect,

- $0.3 \leq d < 0.5$ as a large effect, and

- $d \geq 0.5$ as a very large effect.

In line with these suggestions, previous studies have used the threshold of $d = 0.1$ to accept necessity hypotheses (e.g., van der Valk, Sumo, Dul, & Schroeder, 2016). Rather than relying on this simple cut-off value, researchers should also evaluate the statistical significance of the necessity effect size using NCA's (approximate) permutation test (Dul, 2021, Chapter 4). Small p values (e.g., p < 0.05) indicate the statistical significance of necessity hypotheses. Note that the necessity effect size of an NCA should not be confused with Cohen's d (Cohen, 1988) or the effect size f^2 used in standard PLS-SEM analyses. Different from the d metric, the f^2 depicts the relative impact of an antecedent construct on an outcome in terms of explanatory power (i.e., the f^2 effect size in PLS-SEM determines the change in R^2 if an antecedent construct is omitted in the model; Hair, Hult, Ringle, & Sarstedt, 2022, Chapter 6).

Considerations When Running an NCA

Since the NCA examines necessity conditions between constructs, we first need to extract construct scores from a previous PLS-SEM analysis. Two considerations are important in this respect. First, we need to decide whether to use standardized or unstandardized construct scores as input for the NCA. While basic PLS-SEM analyses generally rely on standardized scores, in an NCA researchers should preferably use unstandardized scores. This is because interpreting the necessity conditions is more intuitive, as it refers to values that correspond to the original (or rescaled) indicator data. In contrast, when we use standardized construct scores, the necessity conditions take values at or close to zero, with some being negative and others being positive. This complicates the interpretation of the necessity conditions. However, using unstandardized construct scores implies that all indicators used in the analysis are measured on the same scale (e.g., on a 7-point Likert scale). To avoid problems that result from using mixed indicator scales (e.g., in combining a 5-point semantic differential scale with a 7-point Likert scale), researchers can also use unstandardized and rescaled scores from a previous IPMA run. This procedure handles problems that result from differently scaled indicator data. Of course, researchers may also run an NCA with standardized data, but this makes the results interpretation less intuitive.

Second, we need to decide whether to use an empirical scale or a theoretical scale as basis for the NCA. The empirical scale refers to the minimum and maximum values as they occur in the dataset. In contrast, the theoretical scale refers to the minimum and maximum values that a construct could take, given the indicator scaling. In most settings, we would use the empirical scale; however, in situations where all indicators have the same theoretical scale (e.g., a 7-point Likert scale) and the empirical data of the resulting construct scores do not make use of the entire scale (e.g., the empirical maximum value is lower than the theoretical maximum value), we would use the theoretical scale.

Results Interpretation

When we interpret the NCA outcomes in PLS-SEM, we can basically expect three kinds of outcomes in terms of a relationship's sufficiency and necessity logic:

1. A substantial and statistically significant relationship (i.e., path coefficient) in the structural model is indeed necessary. Here, the NCA results can help the researcher to determine the minimum level of the construct required to obtain a certain outcome level of a selected target construct.

2. A small and statistically nonsignificant relationship (i.e., path coefficient) in the structural model is necessary. This is an important finding since, even though such relationships are theoretically plausible, researchers often exclude them from their results discussions. The NCA results provide an additional justification to keep such statistically nonsignificant but necessary relationships included in the theoretical model, so that they are considered in the interpretation of results. The NCA results can help the researcher determine minimum levels of the construct required to obtain a certain outcome level of a selected target construct.

3. A small and statistically nonsignificant relationship (i.e., path coefficient) in the structural model is not necessary. In such cases, the researcher can review the theory and the specific data situation (e.g., an analysis for a certain industry or group of people that renders a relationship irrelevant) to improve the model.

In this results interpretation, the necessity logic of NCA complements the sufficiency logic of PLS-SEM in analyzing theoretically established PLS path models and evaluating empirical results. Especially regarding nonsignificant relationships in the path model, the NCA provides important additional results that can guide researchers in their interpretations and recommendations.

Case Study Illustration

We illustrate the joint application of NCA and PLS-SEM using the corporate reputation model. Our illustration focuses on the partial regression in the structural model with the dependent construct customer loyalty (*CUSL*) and its independent constructs competence (*COMP*), customer satisfaction (*CUSA*), and likeability (*LIKE*). We assume that the model has been established as theoretically plausible, that the data are suitable, and that the PLS-SEM results meet the established evaluation criteria. Hair, Hult, Ringle, and Sarstedt (2022, Chapters 4–6) and Sarstedt, Ringle, and Hair (2021) provide further details and results emphasizing that these requirements have been met. We assume that *CUSA* is a precondition for *CUSL* (i.e., there will be no customer loyalty without customer satisfaction). Other explanators of *CUSL*—such as the corporate reputation dimensions *COMP* and *LIKE*—cannot compensate for the absence of customer satisfaction. The NCA allows us to identify the necessary degree of *CUSA* to achieve a certain level of *CUSL*.

All indicators in the corporate reputation model example use an interval scale from 1 (*fully disagree*) to 7 (*fully agree*). Based on the PLS-SEM results, we extract unstandardized construct scores. When the obtained construct scores do not meet the expected minimum and maximum values of 1 and 7 in this example, we use their theoretical scale (i.e., from 1 to 7). Go to the **Example – corporate reputation (advanced)** project, open the **Corporate reputation model**, click on **Calculate**, and select the **PLS-SEM algorithm**. Instead of using the default settings, we change the type of results from **Standardized** to **Unstandardized**. Also, make sure to check the box next to **Open report** (directly above the **Start calculation** button) and click on the **Start calculation** button. After the algorithm has converged, the **Results report** opens, which contains the construct **Scores** under **Final results → Latent variables**. In the toolbar, click on the symbol with the label **Create data file**. In the dialog that follows, change the text under **File name** to **NCA LV scores unstandardized (original scale)** and—under the **Select columns** option—check only the box next to **Latent variable scores**. Next, click on the **Create** button and return to the SmartPLS **Workspace** view (i.e., first click on the arrow symbol with the label **Edit** in the menu bar to return to the **Modeling window**, then click on the arrow symbol with the label **Back** in the menu bar to return to **Workspace** view). The new dataset labeled **NCA LV scores unstandardized (original scale)** now appears in the project **Example – corporate reputation (advanced)**. Double click on the dataset to open the **Data view** and click on the button labeled **Setup** in the menu bar. The dialog that opens includes the eight latent variable (LV) scores variables (i.e., *LV scores*

ATTR to *LV scores QUAL*), the number of missing values (i.e., the value is zero for all variables) and the **Min** and **Max** value. Click on the **Bulk change** option and select 1 as **Min** and 7 as **Max** since we know that the theoretical scale of the constructs is from 1 to 7 in this example (i.e., since we use unstandardized results and all indicators use an interval scale from 1 to 7). Next, click on **Update** to find the **Scale Min** and **Scale Max** has changed to 1 to 7 for all variables in the dataset (i.e., the unstandardized construct scores). Finally, click on the button with the label **Back** in the menu bar and return to the **Workspace** view.

In the SmartPLS **Workspace**, right-click on the **Example – corporate reputation (advanced)** project and choose the **New REGRESSION model** option from the dialog that opens. Alternatively, select the **Example – corporate reputation (advanced)** project and click on the button labeled **Regression** in the bar. A dialog opens in which we enter **NCA for CUSL** under **File name** before clicking on the **Save** button. In the SmartPLS **Workspace** view, a new regression model with the symbol **RM** and the name **NCA for PLS-SEM** in the project appears in the **Example – corporate reputation (advanced)** project. Double click on this model to enter the **Modeling window**. Next, select the **NCA LV scores unstandardized (original scale)** dataset by clicking on the **Select dataset** button, which is displayed above the **Indicators list** on the left-hand side of the **Modeling window**. The LV scores appear in the list of indicators on the left. Preferably, focus the NCA on the target construct *CUSL*. This is done by dragging and dropping the **LV scores – CUSL** on the **Modeling window**. Next, select the **LV scores – COMP, LV scores – CUSA**, and **LV scores – LIKE** indicators and drag-and-drop them on the **LV scores – CUSL** in the **Modeling window**. Our model should look like the one shown in Exhibit 4.5.

To run the NCA for the model shown in Exhibit 4.5, click on the wheel symbol labeled **Calculate** in the tool bar and select the **Necessary condition analysis**

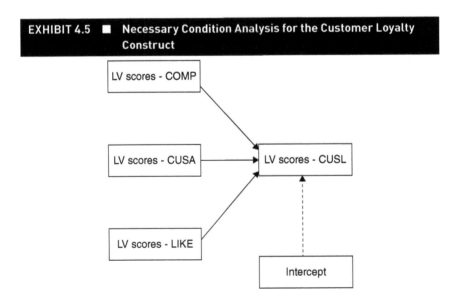

EXHIBIT 4.5 ■ Necessary Condition Analysis for the Customer Loyalty Construct

(NCA) option. Alternatively, we can go to **Calculate → Necessary condition analysis (NCA)** in the menu. A dialog box opens that shows the **Name** of the variables in the model, their **Scale min** of 1, and their **Scale max** of 7—as specified in the previous step. The dialog box also shows the **Observed min**, and their **Observed max** (i.e., the actual minimum and maximum values as present in the dataset). The **Number of steps for bottleneck tables** option enables increasing or decreasing the number of entries in the bottleneck table. When retaining the default value of 10, we start with 0% and have a difference of 10 percentage points between the rows in the bottleneck table (i.e., 11 results rows after the header). In contrast, a value of 20 would increase the number of rows with a distance of 5 percentage points, while 5 would reduce the number of rows to 5 with a distance of 20 percentage points. Keep the default value of 10. Next, ensure that the box next to **Open report** has been ticked and click on **Start calculation** to initiate the analysis.

In the **Results report** that opens, go to **Final results** and click on **Ceiling effect size overview**. The table shown in Exhibit 4.6 appears. For both ceiling lines (i.e., CE-FDH and CR-FDH), most of the necessity effect sizes d are above the minimum value of 0.1, which suggests at least medium effects. The only exception is *LIKE*'s effect size, which is clearly below 0.1 and, thus, does not meet the minimum level to support the *LIKE*'s necessity hypothesis.

The permutation test allows us to test the effect sizes' significances. To initiate this analysis, click on the arrow symbol with the label **Edit** in the menu bar to return to the **Modeling window**, which shows the NCA model (Exhibit 4.5). Next, click on **Calculate → NCA permutation** in the menu and initiate the analysis with the default settings (i.e., **5,000 Permutations**, a **Significance level** of **0.05**, and a **Fixed seed**) by clicking on **Start calculation** (make sure to check the box next to **Open report**). In the **Results report** that opens, go to **Final results** and click on **Ceiling line effect overview**. The resulting table shows the **Original effect size** and the corresponding **95.0%** percentile, which is the upper limit of the 95% one-tailed confidence interval of the necessity effect size d (i.e., upper one-tailed test, since the lowest possible value of the NCA effect size is zero). The table also shows the **Permutation p value** for the CR-FDH ceiling line, which we focus on in our analysis (Exhibit 4.7). All effect sizes are significant as the **Original effect size** is larger than the **95%** percentile. This result is also supported by the **Permutation p value,** which is lower than 0.05 for all variables and both ceiling line types.

EXHIBIT 4.6 ■ Ceiling Effect Size Overview		
	CE-FDH	**CR-FDH**
LV scores - COMP	0.172	0.152
LV scores - CUSA	0.267	0.222
LV scores - LIKE	0.044	0.036

EXHIBIT 4.7 ■ CR-FDH Ceiling Effect Size Significance

Ceiling line effect size overview - CR-FDH   Zoom (100%)   Copy to Excel   Copy to R

	Original effect size	95.0%	Permutation p value
LV scores - COMP	0.152	0.114	0.001
LV scores - CUSA	0.222	0.057	0.000
LV scores - LIKE	0.036	0.010	0.000

The results provide support for the necessity hypotheses relating to the impact of *COMP* and *CUSA* on *CUSL* (i.e., *COMP* and *CUSA* are necessary conditions for *CUSL*). However, although the effect size of *LIKE* is significant, we do not further consider this condition in the analysis since its effect size is small (i.e., $d < 0.1$). We continue the analysis with *CUSA* and *COMP*, which have medium and statistically significant necessity effect sizes.

Next, we return to the NCA results in SmartPLS. To do so, leave the NCA permutation **Results report** by clicking on the arrow symbol with the label **Edit** in the menu bar. In the **Modeling window** that appears, select the NCA results (above the model, under **Report**) and click on the **Open report** button. Alternatively, simply rerun the NCA computations. Under **Final results** → **NCA charts**, the ceiling lines can be graphically inspected. Exhibit 4.8 shows the ceiling line chart for the relationship between *COMP* and *CUSL*. When inspecting the CR-FDH ceiling line, we find, for example, that *COMP* must have a value of at least 1.912 for *CUSL* to take a value of 4. The area without observations (i.e., the ceiling zone C) and the area with observations (i.e., the scope S) determine *COMP*'s necessity effect size (i.e., the larger the area C, the larger the effect size). Exhibit 4.9 shows the ceiling line chart for the relationship between *CUSA* and *CUSL*. The vertical alignment of the observations in this chart results from the fact that *CUSA* is a single-item construct. The responses of the *cusa* indicator on an interval scale from 1 to 7 become the *CUSA*'s construct scores. Note that the ceiling zone C in the chart for *CUSA* on *CUSL* is larger than in the chart for *COMP* on *CUSL* (Exhibit 4.8). This corresponds with *CUSA*'s larger necessity effect size.

In the next step, we extract the bottleneck tables, focusing on the CR-FDH results. Under **Final results**, we therefore go to **Bottleneck tables – CR-FDH** → **Values**. Exhibit 4.10 shows the **Bottleneck tables – CR-FDH** → **Values** results table, which has 11 rows below the header. The first row starts with a level of 0% for the dependent construct, which corresponds with *CUSL*'s minimum level of 1. On the other extreme, 100% represents *CUSL*'s maximum value of 7, while 50% represents the mid-level of 4 on *CUSL*'s unstandardized scale from 1 to 7. The CR-FDH bottleneck table shows the necessary threshold levels of *COMP* and *CUSA*, which are required to achieve a certain *CUSL* value. We focus on these two constructs, which are necessary conditions since, at least, they have a medium effect size, which is significant. For example, to produce a 50% level of *CUSL*, *COMP* must have a value of at least 1.912 and *CUSA* a value of at

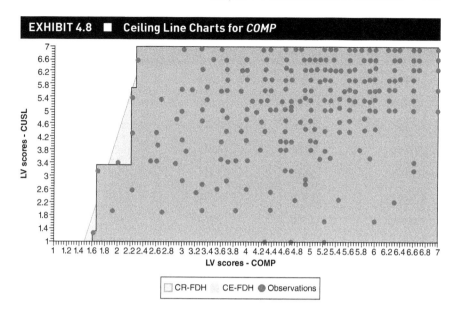

EXHIBIT 4.8 ■ Ceiling Line Charts for *COMP*

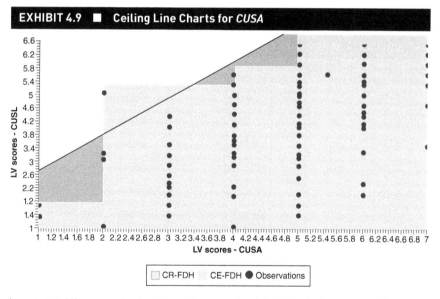

EXHIBIT 4.9 ■ Ceiling Line Charts for *CUSA*

least 2.101. The example also shows that up to level 2.200, which corresponds to 20% of *CUSL*'s performance, *CUSA* is not necessary (NN).

Besides the previously analyzed **Values** results table, the bottleneck tables can also be displayed in terms of **Counts** and **Percentiles** (Exhibit 4.11). This results table shows the number of observations (i.e., obtained from the **Counts** results in SmartPLS table) below a certain threshold value and gives (in brackets) the related information in percentage terms (obtained from the **Percentiles** results table in SmartPLS). These additional

EXHIBIT 4.10 ■ CR-FDH Values Bottleneck Table

	LV scores – CUSL	*LV scores – COMP*	*LV scores – CUSA*	*LV scores – LIKE*
0.000%	1.000	1.471	NN	NN
10.000%	1.600	1.559	NN	NN
20.000%	2.200	1.647	NN	NN
30.000%	2.800	1.735	1.031	NN
40.000%	3.400	1.824	1.566	NN
50.000%	4.000	1.912	2.101	NN
60.000%	4.600	2.000	2.637	NN
70.000%	5.200	2.088	3.172	1.053
80.000%	5.800	2.177	3.707	1.491
90.000%	6.400	2.265	4.243	1.928
100.000%	7.000	2.353	4.778	2.366

EXHIBIT 4.11 ■ CR-FDH Counts and Percentiles Bottleneck Table

LV scores – CUSL		*LV scores – COMP*	*LV scores – CUSA*	*LV scores – LIKE*
0.0%	1.0	0 (0%)	0 (0%)	0 (0%)
10.0%	1.6	0 (0%)	0 (0%)	0 (0%)
20.0%	2.2	1 (0.291%)	0 (0%)	0 (0%)
30.0%	2.8	2 (0.581%)	2 (0.581%)	0 (0%)
40.0%	3.4	2 (0.581%)	2 (0.581%)	0 (0%)
50.0%	4.0	2 (0.581%)	6 (1.744%)	0 (0%)
60.0%	4.6	4 (1.163%)	6 (1.744%)	0 (0%)
70.0%	5.2	4 (1.163%)	21 (6.105%)	6 (1.744%)
80.0%	5.8	4 (1.163%)	21 (6.105%)	9 (2.616%)
90.0%	6.4	7 (2.035%)	62 (18.023%)	19 (5.523%)
100.0%	7.0	8 (2.326%)	62 (18.023%)	39 (11.337%)

Note: The values given in brackets indicate the percentage of observations below the threshold level of the independent constructs, that would be required to obtain a certain outcome level of the dependent variable.

outcomes indicate how difficult it is to reach a certain outcome level. For example, assuming a CUSL level of 90%, 62 of the 344 observations in the dataset (18.023%) do not support this desired outcome in terms of *CUSA*. This percentage is considerably lower for *COMP*.

In a final step, we contrast the NCA results with those of the standard PLS-SEM analysis. Running the standard PLS-SEM algorithm shows that *CUSA* has the strongest effect on *CUSL*, followed by *LIKE*. In contrast, *COMP* has a non-significant effect on *CUSL*, which is very close to zero. The NCA, on the other hand, shows that *COMP* is a necessary condition—as is *CUSA*. Therefore, the strong path coefficients of *CUSA* and *LIKE* can only be used to achieve a certain outcome level of *CUSL* if certain minimum levels of the necessary conditions *COMP* and *CUSA* are achieved. By contrasting these results, we can showcase the NCA's ability to complement the PLS-SEM estimation results of the corporate reputation model.

IMPORTANCE-PERFORMANCE MAP ANALYSIS

Overview

The **IPMA** (also called **importance-performance matrix**, **impact-performance map**, or **priority map analysis**) extends the standard PLS-SEM results reporting of path coefficient estimates and other parameters by adding a procedure that considers the average values of the construct scores (e.g., Magno & Dossena, 2023; Ringle & Sarstedt, 2016; Streukens, Leroi-Werelds, & Willems, 2017). More precisely, the IPMA contrasts the total effects, representing the predecessor constructs' **importance** in predicting a specific target construct, with their average construct scores indicating their **performance**. The rational is to identify predecessor constructs that have a relatively high importance for explaining the target construct (i.e., those that have a strong **total effect**), but also have a relatively low performance (i.e., low average construct scores), so that improvements can be implemented. The following descriptions draw on Ringle and Sarstedt's (2016) article on the method.

To illustrate the concept of the IPMA, consider the path model in Exhibit 4.12 with four constructs Y_1 to Y_4. In this path model, Y_4 represents the final target construct, directly predicted by Y_1, Y_2, and Y_3. Furthermore, Y_2 has an **indirect effect** on Y_4 via Y_3, while Y_1 has an indirect effect via both Y_2 and Y_3. Adding the predecessor constructs' direct and indirect effects yields their total effects on Y_4, which represent the importance dimension in the IPMA. In contrast, these constructs' average scores—indicated by the numbers in the constructs—represent their performance, in the sense that high values indicate a greater performance.

The IPMA combines these two aspects graphically by relating the unstandardized total effects on the *x*-axis with the construct scores, rescaled on a range from 0 to 100, on the *y*-axis. The result is a chart, such as the one in Exhibit 4.13. For interpretation, we focus on constructs in the lower right area of the importance-performance map. These constructs have a high importance for the target construct, such that they have a strong

EXHIBIT 4.12 ■ Importance-Performance Map Analysis Example

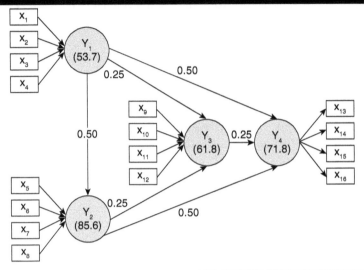

Source: Adapted from Ringle & Sarstedt (2016).

EXHIBIT 4.13 ■ Importance-Performance Map of the Target Construct Y_4

Source: Adapted from Ringle & Sarstedt (2016).

impact but show a low performance. Consequently, there is a particularly high potential to improve the performance of the constructs positioned in this area. Conversely, the performance improvements priority of constructs with lower importance relative to the others is less pronounced. In Exhibit 4.13, Y_1 is particularly important for explaining the target construct Y_4. More precisely, ceteris paribus (i.e., when everything else remains constant), increasing Y_1's performance by one unit will increase the performance of Y_4 by the value of Y_1's total effect on Y_4, which is 0.84. Since the performance of Y_1 is relatively low, there is substantial room for improvement, underscoring the importance of addressing the underlying aspects of this construct and, hence, rendering it especially pertinent for managerial

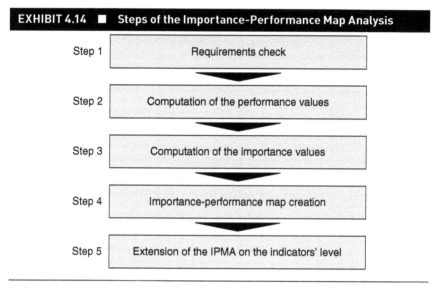

EXHIBIT 4.14 ■ Steps of the Importance-Performance Map Analysis

Step 1 — Requirements check

Step 2 — Computation of the performance values

Step 3 — Computation of the importance values

Step 4 — Importance-performance map creation

Step 5 — Extension of the IPMA on the indicators' level

Source: Adapted from Ringle & Sarstedt (2016).

or policy interventions. While this introductory example illustrates the IPMA on the construct level, the analysis can also be run on the indicator level. In this case, individual data points in the importance-performance map are derived from indicator mean values and their total effect on a particular target construct.

Several PLS-SEM studies underline the IPMA results' usefulness for offering important insights into the predecessor constructs' role and their relevance for managerial actions (e.g., Höck, Ringle, & Sarstedt, 2010; Kristensen, Martensen, & Grønholdt, 2000; Magno & Dossena, 2023; Zabukovšek, Bobek, Zabukovšek, Kalinic, & Tominc, 2022). The IPMA is also particularly useful when contrasting PLS-SEM results from a multigroup analysis (Chapter 5), as several studies illustrate (Rigdon, Ringle, Sarstedt, & Gudergan, 2011; Schloderer, Sarstedt, & Ringle, 2014; Völckner, Sattler, Hennig-Thurau, & Ringle, 2010).

The IPMA draws on the five-step procedure shown in Exhibit 4.14 (Ringle & Sarstedt, 2016). The first step involves checking whether the requirements for carrying out the analysis have been fulfilled (Step 1). The analysis proceeds with the computation of the constructs' performance values (Step 2) and their importance values (Step 3). The importance-performance map creation for a selected target construct is based on these results (Step 4). Finally, the IPMA can be extended to the indicator level to obtain more specific information on those managerial or policy actions that can be most effective (Step 5). The following sections explain each step in greater detail.

Systematic IPMA Execution

Step 1: Requirements Check

IPMA applications must meet three requirements. First, the **rescaling** of the construct scores on a range from 0 to 100 requires all indicators in the PLS path model to have a

metric scale or at least an **equidistant scale**. In an equidistant scale, the intervals are distributed in equal units, which is typically the case when using an **ordinal scale**. However, in order for an ordinal scale to approximate being equidistant, the scale needs to be balanced, which means there is an equal number of positive and negative categories. Therefore, a **forced-choice scale** (i.e., scales without a neutral category) is, strictly speaking, not equidistant. On the contrary, a 5-point Likert scale with two negative categories (completely disagree and disagree), a neutral option, and two positive categories (agree and completely agree) can be considered an equidistant scale. Indicators measured on a **nominal scale**, however, cannot be used in an IPMA. In a nominal scale, the assignments of numbers to certain object characteristics are interchangeable, rendering any rescaling arbitrary.

Second, all indicator coding must have the same scale direction. A low value on the scale must represent a negative or low outcome, and a high value must represent a positive or high outcome. If this requirement is not met, we cannot conclude that higher construct scores represent better performance. In this case, the respective indicator coding needs to be changed by reversing the scale (e.g., on a 5-point scale, 5 becomes 1 and 1 becomes 5, 4 becomes 2 and 2 becomes 4, and 3 remains unchanged).

Third, regardless of whether the measurement model is specified reflectively or formatively, the outer weights (i.e., indicator weights) estimates must be positive. If the outer weights are negative, the construct scores will not fall within the 0 to 100 range, but would, for example, be between –5 and 95. Note that there are different reasons for (unexpected) negative outer weights. If an outer weight is negative and significant, the researcher should inspect the indicator and its scale. It may have another direction compared to the other indicators in the measurement model, which requires reversing the scale. In case of nonsignificant outer weights (with negative signs), the researcher should consider removing those indicators. Finally, negative outer weights might be a result of high indicator collinearity. For example, variance inflation factor (VIF) values of 3 and higher indicate a potential collinearity problem (Hair, Hult, Ringle, & Sarstedt, 2022, Chapter 5). In this case, the researcher should also consider removing indicators. However, removing indicators from measurement models involves some additional considerations as explained by Hair, Hult, Ringle, and Sarstedt (2022, Chapter 5) in greater detail.

While not being a formal requirement for running an IPMA, researchers should carefully consider PLS path model setups that favor the IPMA use on the indicator level. When a construct of high priority for a specific target construct is identified, it is particularly advantageous to further analyze this predecessor construct's measurement model on the indicator level. Such an assessment is especially useful when the measurement model is specified as formative (see Chapter 1)—as illustrated in our sample model in Exhibit 4.12 (i.e., indicators x_1 to x_{12}). In this case, the indicators describe aspects that shape the corresponding construct, while their weights indicate each aspect's importance in this regard. Therefore, aspects underlying indicators with high weights should be given more attention to identify managerial or policy actions aimed at improving the target construct's performance. Note that the IPMA can be applied on any kind of PLS path model, regardless of whether the constructs' measurement models are formative or reflective. The IPMA builds on the outer weights—as explained in more detail in the subsequent sections—and PLS-SEM always provides outer weights estimates, also when a measurement model is specified as reflective.

Step 2: Computation of the Performance Values

The indicator data determine the construct scores and, thus, their performance. Similarly, when conducting an IPMA on the indicator level, the mean value of an indicator represents its average performance. When computing average values on the construct or indicator level, it is important to remember that indicators might be measured on different scales. For example, some indicators might use a scale with values from 1 to 5, while others use a scale with values from 1 to 7 or from 1 to 9. To facilitate the interpretation and comparison of performance levels, the IPMA rescales indicator scores on a range between 0 and 100, with 0 representing the lowest and 100 representing the highest performance. Since most researchers are familiar with interpreting percentage values, this kind of performance scale is easy to understand. The rescaling of an observation j with respect to indicator i proceeds via

$$x_{ij}^{rescaled} = \frac{E[x_{ij}] - min[x_i]}{max[x_i] - min[x_i]} \cdot 100,$$

where the i-th indicator in the PLS path model, E[.] represents indicator i's actual score of respondent j, and min[.] and max[.] represent the indicator's minimum and maximum value. It is important to note that the minimum and maximum values refer to the potential values on a certain scale (e.g., 1 and 5 on a 1 to 5 scale) and not the minimum and maximum values of the actual responses (e.g., 2 and 4 on a 1 to 5 scale). For example, according to this formula, a value of 4 on a 1 to 5 scale becomes. $(4 - 1)/(5 - 1) \cdot 100 = 75$ All data points used for estimating the PLS path model are rescaled this way.

Exhibit 4.15 shows an excerpt of the original indicator data ($N = 300$) used to estimate the sample model from Exhibit 4.12. All indicators are measured on a scale from 1 to 5. Moreover, Exhibit 4.16 shows the indicator data from Exhibit 4.15, rescaled on a range from 0 to 100, which serve as input for the computation of the rescaled construct scores. In addition, the mean values of the rescaled indicators represent their performance values (e.g., 79 for indicator x_1 and 77.5 for indicator x_2), which are later used for the IPMA on the indicator level.

Irrespective of whether the measurement model of a construct is reflective (i.e., Y_4) or formative (i.e., Y_1, Y_2, and Y_3), the rescaled construct scores are a linear combination of the rescaled indicator data and the rescaled outer weights. Hence, the IPMA can be executed for reflective and formative constructs but uses only the outer weights in both

EXHIBIT 4.15 ■ Original Indicator Data

Case	X_1	X_2	X_3	X_4	X_5	X_6	X_7	X_8	X_9	X_{10}	X_{11}	X_{12}	X_{13}	X_{14}	X_{15}	X_{16}
1	5	2	1	5	5	2	4	3	3	3	4	5	1	2	4	3
2	4	5	2	3	4	1	1	2	5	4	5	3	3	5	4	3
3	4	1	3	4	5	3	5	3	5	3	4	5	3	5	3	1
4	1	3	2	1	1	4	1	3	3	2	2	4	5	1	4	4
5	3	2	1	3	5	2	2	3	5	1	2	3	1	3	4	5
...	...	...	...	...	...	...	...	...	...	...	...	...	...	...	...	...
299	2	4	3	4	4	2	2	1	4	3	3	4	2	5	4	3
300	2	4	3	1	1	1	4	3	5	2	3	5	4	2	1	5
Mean Value	4.2	4.1	3.4	2.3	3.6	4.5	3.4	3.1	3.4	4.7	4.4	4.6	3.4	4.2	4.5	4.2

Source: Adapted from Ringle & Sarstedt (2016).

EXHIBIT 4.16 ■ **Rescaled Indicator Data**

Case	X_1	X_2	X_3	X_4	X_5	X_6	X_7	X_8	X_9	X_{10}	X_{11}	X_{12}	X_{13}	X_{14}	X_{15}	X_{16}
1	100	25	0	100	100	25	75	50	50	50	75	100	0	25	75	50
2	75	100	25	50	75	0	0	25	100	75	100	50	50	100	75	50
3	75	0	50	75	100	50	100	50	100	50	75	100	50	100	50	0
4	0	50	25	0	0	75	0	50	50	25	25	75	100	0	75	75
5	50	25	0	50	100	25	25	50	100	0	25	50	0	50	75	100
...	...	...	...	...	...	...	...	...	...	...	...	...	...	...	...	...
299	25	75	50	75	75	25	25	0	75	50	50	75	25	100	75	50
300	25	75	50	0	0	0	75	50	100	25	50	100	75	25	0	100
Mean Value	79.0	77.5	59.5	33.5	66.0	87.5	59.5	53.0	59.5	91.5	85.5	90.0	59.5	79.0	88.5	79.0

Source: Adapted from Ringle & Sarstedt (2016).

cases. Even though the IPMA permits using both types of measurement models (as well as single item constructs), it is particularly useful in situations where the endogenous target construct has a reflective measurement model, while the exogenous constructs have formative measurement models. In this situation, the performance improvements of indicators in the formative measurement models result in a performance increase of the exogenous constructs, which in turn translates into a performance enhancement of the key target construct.

To obtain the rescaled weights, we must first compute the unstandardized weights by dividing each standardized weight by the standard deviation of its respective indicator. While the standardized outer weights originate from the standard PLS path model estimation, the estimation of each indicator's standard deviation is based on the original indicator data. For example, if x_1 has a standardized weight of 0.2 and a standard deviation of 1.619, the resulting unstandardized weight is 0.124.

Finally, we rescale the unstandardized outer weights, so that their sum equals one per measurement model. For this purpose, we need to divide each indicator's unstandardized weight (e.g., 0.124 for indicator x_1) by the sum of the unstandardized weights of all the indicators that belong to the same measurement model. For Y_1, the sum of all the unstandardized indicator weights is 0.124 + 0.168 + 0.191 + 0.406 = 0.889. Therefore, for indicator x_1, we obtain the unstandardized and rescaled outer weight of 0.139 after dividing 0.124 by 0.889. Exhibit 4.17 shows the standardized, unstandardized, and rescaled outer weights along with the indicators' standard deviations with regard to our sample model in Exhibit 4.12.

In the next step, the IPMA uses the rescaled indicator data (Exhibit 4.16) and the rescaled outer weights (Exhibit 4.17) to compute the rescaled construct scores by means of simple linear combinations. For example, the first data point in the vector of Y_1's scores is

$$100 \cdot 0.139 + 25 \cdot 0.189 + 0 \cdot 0.215 + 100 \cdot 0.457 \approx 64.3$$

Exhibit 4.18 shows the resulting construct scores along with their mean values. In our example, Y_1 has a mean value (i.e., performance) of 53.7, Y_2 of 85.6, Y_3 of 61.8, and Y_4 of 78.1. These results serve as input for the importance-performance map's performance dimension.

Step 3: Computation of the Importance Values

A construct's importance in terms of explaining another directly or indirectly linked (target) construct in the structural model is derived from the total effect of the relationship between these two constructs. The total effect is the sum of the direct and all the indirect effects in the structural model (Hair, Hult, Ringle, & Sarstedt, 2022, Chapter 6). For example, to determine the total effect of Y_1 on Y_4 (Exhibit 4.12), we need to consider the **direct effect** of the relationship between these two constructs (0.50) and the following three indirect effects via Y_2 and Y_3, respectively.

EXHIBIT 4.17 ■ Computation of Unstandardized and Rescaled Outer Weights

Latent variable	Indicator	Standardized outer weights	Standard deviation of the indicators	Unstandardized outer weights	Rescaled outer weights
Y_1	x_1	0.2	1.619	0.124	0.139
	x_2	0.3	1.789	0.168	0.189
	x_3	0.4	2.099	0.191	0.215
	x_4	0.5	1.231	0.406	0.457
Y_2	x_5	0.1	2.099	0.048	0.114
	x_6	0.1	1.101	0.091	0.218
	x_7	0.4	3.357	0.119	0.285
	x_8	0.4	2.504	0.160	0.383

EXHIBIT 4.17 ■ Computation of Unstandardized and Rescaled Outer Weights *(Continued)*

Latent variable	Indicator	Standardized outer weights	Standard deviation of the indicators	Unstandardized outer weights	Rescaled outer weights
Y_3	x_9	0.1	1.762	0.057	0.136
	x_{10}	0.1	1.744	0.057	0.137
	x_{11}	0.4	2.270	0.176	0.422
	x_{12}	0.4	2.653	0.151	0.361
Y_4	x_{13}	0.3	2.164	0.139	0.186
	x_{14}	0.3	1.874	0.160	0.215
	x_{15}	0.3	1.413	0.212	0.286
	x_{16}	0.3	1.291	0.232	0.313

Source: Adapted from Ringle & Sarstedt (2016).

EXHIBIT 4.18 ■ Rescaled Construct Scores				
Case	Y_1	Y_2	Y_3	Y_4
1	64.3	76.3	63.3	42.5
2	57.6	75.4	18.2	67.9
3	55.5	82.0	76.5	45.1
4	14.8	47.0	26.9	63.5
5	34.5	37.7	43.3	63.5
...	...	...	...	...
299	62.7	62.4	22.9	63.3
300	28.4	69.4	47.1	50.6
Mean Value	53.7	85.6	61.8	78.1

Source: Adapted from Ringle & Sarstedt (2016).

$$Y_1 \rightarrow Y_2 \rightarrow Y_4 = 0.50 \bullet 0.50 = 0.25$$
$$Y_1 \rightarrow Y_2 \rightarrow Y_3 \rightarrow Y_4 = 0.50 \bullet 0.25 \bullet 0.25 = 0.03125$$
$$Y_1 \rightarrow Y_3 \rightarrow Y_4 = 0.25 \bullet 0.25 = 0.0625$$

Adding up the individual indirect effects yields the total indirect effect of Y_1 on Y_4, which is approximately 0.34. Therefore, the total effect of Y_1 on Y_4 is 0.84 (= 0.50 + 0.34). This total effect expresses the importance in predicting the target construct of Y_4.

It is noteworthy to stress that the IPMA draws on unstandardized effects to facilitate a ceteris paribus (i.e., when everything else remains constant) interpretation of the impact of predecessor constructs on the target construct (i.e., analogous to the interpretation of unstandardized regression coefficients; Hair, Black, Babin, & Anderson, 2018, Chapter 5). More precisely, by drawing on unstandardized effects, we can conclude that an increase in a certain predecessor construct's performance by one performance point would increase the target construct's performance points by the size of the unstandardized total effect of this predecessor construct on the target construct. Alternatively, we could use standardized total effects. However, the interpretation of standardized path coefficients and total effects can be somewhat less intuitive compared to their unstandardized versions. A one standard deviation change in the predecessor construct is associated with a change in the target construct in standard deviation values equal to the size of the standardized total effect of the predecessor construct on the target construct. Hence, we continue the further illustrations in this chapter with reference to the unstandardized PLS-SEM path coefficients and total effects.

EXHIBIT 4.19   ■   Unstandardized Direct, Indirect, and Total Effects in the IPMA

Predecessor Construct	Direct Effect on Y_4	Indirect Effect on Y_4	Total Effect on Y_4
Y_1	0.50	0.34	0.84
Y_2	0.50	0.06	0.56
Y_3	0.25	—	0.25

Source: Adapted from Ringle & Sarstedt (2016).

Note: All effects denote unstandardized effects.

Exhibit 4.19 summarizes all the unstandardized total effects with respect to our target construct of Y_4. Note that Y_3 does not have an indirect effect on Y_4. Therefore, its total effect equals the direct effect of 0.25. At this point, after computing the importance and performance values, all information required to draw the importance-performance map is available.

Step 4: Importance-Performance Map Creation

The IPMA focuses on one key target construct of interest in the PLS path model. Therefore, the first step in creating an importance-performance map requires selecting the target construct. In our example in Exhibit 4.12, Y_4 represents such a key target construct. To create the importance-performance map of Y_4, we need to use the importance and performance values of Y_4's predecessor constructs (i.e., Y_1, Y_2, and Y_3). Exhibit 4.20 summarizes the values of this map's importance and performance dimensions, as obtained in the previous IPMA steps.

Drawing a scatter plot using the information shown in Exhibit 4.20 enables us to create an importance-performance map as shown in Exhibit 4.13. The x-axis represents the importance of Y_1, Y_2, and Y_3 for explaining the target construct Y_4, while the y-axis depicts the performance of Y_1, Y_2, and Y_3 in terms of their average rescaled construct scores. For a better orientation, researchers can also draw two additional lines in the importance-performance map: the mean importance value (i.e., a vertical line) and the mean performance value (i.e., a horizontal line) of the displayed constructs (Exhibit 4.21). With regard to our example Y_1, Y_2, and Y_3 have a mean importance of 0.55 and a mean performance of 67.0 (Exhibit 4.20). These two additional lines divide the importance-performance map into four areas with importance and performance values below and above the average. Generally, when analyzing the importance-performance map, constructs in the lower right area (i.e., above-average importance and below-average performance) represent the greatest opportunity to achieve improvement, followed by the upper right, lower left, and, finally, the upper left areas. Thereby, the importance-performance map provides guidance for the prioritization of managerial or policy activities of high importance for the aspect underlying the selected target, but which require performance improvements.

EXHIBIT 4.20   ■   Data of the Importance-Performance Map for Construct Y_4

Predecessor Construct	Importance	Performance
Y_1	0.84	53.7
Y_2	0.56	85.6
Y_3	0.25	61.8
Mean Value	0.55	67.0

Source: Adapted from Ringle & Sarstedt (2016).

EXHIBIT 4.21   ■   Adjusted Importance-Performance Map of Target Construct Y_4

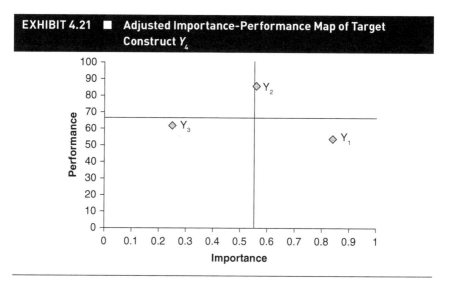

Source: Adapted from Ringle & Sarstedt (2016).

In our example, the importance-performance map (Exhibit 4.21) shows that Y_1 has a relatively low performance of 53.7. In comparison to the other constructs, the performance is below average. On the other hand, with a total effect of 0.84, this construct's importance is particularly high. Therefore, a one-unit increase in Y_1's performance from 53.7 to 54.7 would increase the performance of Y_4 by 0.84 points from 78.10 to 78.94. Hence, when the objective of managers is increasing the performance of the target construct, Y_4, their first priority should be to improve the performance of aspects captured by Y_1 since this construct has the highest (above average) importance, but a relatively low (below average) performance. Aspects related to constructs Y_2 and Y_3 follow as a second and third priority.

Step 5: Extension of the IPMA to the Indicator Level

The IPMA is not limited to the construct level. We can also conduct an IPMA on the indicator level to identify relevant and even more specific areas of improvement. More

EXHIBIT 4.22 ■ **Adjusted Importance-Performance Map of Y_1, Y_2, and Y_3's Indicators on Target Construct Y_4**

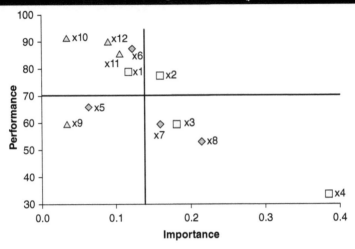

Source: Adapted from Ringle & Sarstedt (2016).

precisely, we can interpret the rescaled outer weights as an indicator's relative importance compared to that of the other indicators in a specific measurement model.

The importance values are derived from the indicators' total effects on the target construct, which is the result of multiplying the rescaled outer weights of a predecessor construct's indicators with its unstandardized total effect on the target construct. For example, with regard to the indicators of Y_1, we would multiply the rescaled outer weights of x_1 to x_4 (i.e., 0.139, 0.189, 0.215, and 0.457; Exhibit 4.17) with the unstandardized total effect of Y_1 on Y_4 in the structural model (i.e., 0.84). This analysis yields importance values of x_1 to x_4 of 0.117, 0.159, 0.181, and 0.384, respectively. The performance values are derived from the indicators' mean value of the rescaled data (i.e., 79.0, 77.5, 59.5, and 33.5; Exhibit 4.16). With this data for all indicators of Y_1, Y_2, and Y_3, we can create an importance-performance map as shown in Exhibit 4.22.

These results suggest indicator x_4 should be given the highest priority for improvement since it has the highest relative importance but the lowest performance. A one-unit point increase in x_4's performance increases the performance of Y_4 by x_4's importance value (i.e., the unstandardized total effect in this example), which is 0.384 (ceteris paribus). Indicators x_8, x_3, x_7, and x_2 follow with second to fifth priority. The other indicators shown in Exhibit 4.22 are less relevant for improving Y_4's performance. Instead of creating a single importance-performance map for all indicators (as shown in Exhibit 4.22), it is often more useful in applications to create importance-performance maps for the indicators of each construct separately (i.e., three indicator-level importance-performance map for each construct Y_1, Y_2, and Y_3 in our example). Exhibit 4.23 summarizes the rules of thumb for conducting the IPMA.

EXHIBIT 4.23 ■ Rules of Thumb for Executing Importance-Performance Map Analyses

- Only use indicators measured on a metric scale or equidistant scale (typically an ordinal or higher scale).

- All indicator coding must have the same scale direction. For example, a low value on the scale must represent a negative or low outcome, and a high value must represent a positive or high outcome. Otherwise, we cannot conclude that higher construct scores represent better performance).

- Regardless of whether the measurement model is specified as formative or reflective, the outer weights estimates must have the same (e.g., positive) sign to ensure interpretability of results.

- For the interpretability of results, the total effects in the structural model should have the same (e.g., positive) signs.

- Select a key target construct and run the IPMA for its direct or all predecessor constructs in the structural model.

- Use the unstandardized total effects and the average construct scores on a scale from 0 to 100 to create the importance-performance map.

- Focus on those constructs with relatively low performance and a relatively high importance (i.e., a large total effect) on the target construct. The ceteris paribus interpretation of results assumes that an increase of a predecessor construct's performance by one point increases the performance of the selected key target construct by the size of the total effect.

Case Study Illustration

To illustrate the IPMA application, we draw on the corporate reputation model, following the procedure depicted in Exhibit 4.14. First, we check the requirements for carrying out an IPMA (Step 1). After reviewing the questionnaire, we find that the indicator data have been measured on a 7-point Likert scale with balanced categories and a neutral category. The indicators are measured on an equidistant scale. Furthermore, for all of the indicators, a higher value represents a better outcome (see Chapter 1). Therefore, we do not need to reverse-scale any of the indicators. To obtain further information on the data, go to the **Workspace** view and double-click on **Complete data [344]** in the **Example – corporate reputation (advanced)** project.

The **Data view** opens (Exhibit 4.24), which provides further information on the dataset (e.g., the number of missing values) and some descriptive statistics for the variables (e.g., the standard deviation). When clicking on the **Setup** button in the toolbar, a window opens that allows us to adjust settings related to the entire dataset (e.g., **Delimiter character, Escape character, Locale, Missing value treatment**) or specific variables (i.e., **Name, Scale, Min, Max**) as shown in Exhibit 4.25. We first need to check that all the variables used the full range of values on the 7-point scale as indicated

EXHIBIT 4.24 ■ SmartPLS Data View

42 indicators with 344 cases and 11 missing values Zoom (100%) ———○——— Copy to Excel

Name	No.	Type	Missings	Mean	Median	Scale min	Scale max	Observed min	Observed max	Standard deviation	Excess kurtosis	Skewness	Cramér-von Mises p value
serviceprovider	1	CAT	0	2.000	2.000	1.000	4.000	1.000	4.000	1.003	-0.513	0.747	0.000
servicetype	2	0|1	0	1.637	2.000	1.000	2.000	1.000	2.000	0.481	-1.684	-0.571	0.000
comp_1	3	MET	0	4.648	5.000	1.000	7.000	1.000	7.000	1.433	-0.324	-0.264	0.000
comp_2	4	MET	0	5.424	6.000	1.000	7.000	1.000	7.000	1.375	-0.616	-0.586	0.000
comp_3	5	MET	0	5.221	6.000	1.000	7.000	1.000	7.000	1.458	-0.188	-0.677	0.000
like_1	6	MET	0	4.584	5.000	1.000	7.000	1.000	7.000	1.547	-0.399	-0.405	0.000
like_2	7	MET	0	4.250	4.000	1.000	7.000	1.000	7.000	1.848	-0.901	-0.312	0.000
like_3	8	MET	0	4.480	5.000	1.000	7.000	1.000	7.000	1.871	-0.941	-0.325	0.000
cusl_1	9	MET	3	5.129	5.000	1.000	7.000	1.000	7.000	1.513	0.268	-0.792	0.000
cusl_2	10	MET	4	5.276	6.000	1.000	7.000	1.000	7.000	1.744	0.040	-0.951	0.000
cusl_3	11	MET	3	5.651	6.000	1.000	7.000	1.000	7.000	1.655	0.930	-1.301	0.000
cusa	12	MET	1	5.440	6.000	1.000	7.000	1.000	7.000	1.174	0.777	-0.768	0.000
csor_1	13	MET	0	4.235	4.000	1.000	7.000	1.000	7.000	1.469	-0.363	-0.042	0.000
csor_2	14	MET	0	3.076	3.000	1.000	7.000	1.000	7.000	1.651	-0.564	0.497	0.000
csor_3	15	MET	0	3.988	4.000	1.000	7.000	1.000	7.000	1.478	-0.468	-0.061	0.000
csor_4	16	MET	0	3.125	3.000	1.000	7.000	1.000	7.000	1.462	-0.463	0.197	0.000
csor_5	17	MET	0	3.983	4.000	1.000	7.000	1.000	7.000	1.583	-0.696	-0.042	0.000
csor_global	18	MET	0	4.988	5.000	1.000	7.000	1.000	7.000	1.289	-0.629	-0.142	0.000
attr_1	19	MET	0	4.991	5.000	1.000	7.000	1.000	7.000	1.458	-0.078	-0.567	0.000
attr_2	20	MET	0	2.945	2.000	1.000	7.000	1.000	7.000	2.098	-1.131	0.574	0.000
attr_3	21	MET	0	4.811	5.000	1.000	7.000	1.000	7.000	1.451	-0.568	-0.275	0.000
attr_global	22	MET	0	5.587	6.000	2.000	7.000	2.000	7.000	1.214	-0.100	-0.654	0.000

EXHIBIT 4.25 ■ SmartPLS Data Setup

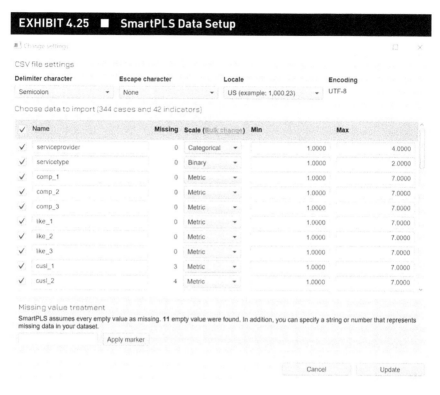

by the minimum and maximum indicator values. SmartPLS automatically reads these minimum and maximum values from the data. If the respondents have not made use of the full scale (e.g., the actual minimum value is 2 instead of 1 or the maximum value is 6 instead of 7), SmartPLS cannot correctly rescale the data on a scale from 0 to 100. If this is the case, change the minimum and/or maximum value manually by entering the correct value under **Min** and/or **Max** in the data **Setup** window. After closing the data **Setup** window, we return to the **Data view** (Exhibit 4.24), which shows the minimum and maximum values after potential correction under **Scale min** and **Scale max**. These values are used for the IPMA analysis.

Next, we inspect the (unstandardized) signs of the outer weights obtained from the PLS-SEM model estimation. To do so, double click on **Corporate reputation model** in the **Example – corporate reputation (advanced)** project (Exhibit 4.24). The corporate reputation model opens in the **Modeling window** and, in the toolbar, we select **Calculate → PLS-SEM algorithm**. In the dialog box, keep the default settings, except for the **Type of results**, which we change into **Unstandardized** by clicking on the corresponding combo box. Also make sure to check the box next to **Open report**. In the results report that opens, go to **Final results** and click on **Outer weights →** **Matrix** to display the indicator weights in a matrix format (Exhibit 4.26). All the outer weights are positive, except for *qual_4*, whose weight is −0.003. However, this indicator weight is very close to zero and nonsignificant. Hence, this indicator's impact on

EXHIBIT 4.26 ■ SmartPLS Results Report

Outer weights - Matrix   Zoom (100%)   [Copy to Excel] [Copy to R]

	ATTR	COMP	CSOR	CUSA	CUSL	LIKE	PERF	QUAL
attr_1	0.341							
attr_2	0.115							
attr_3	0.544							
comp_1		0.386						
comp_2		0.313						
comp_3		0.301						
csor_1			0.253					
csor_2			0.027					
csor_3			0.334					
csor_4			0.067					
csor_5			0.319					
cusa				1.000				
cusl_1					0.345			
cusl_2					0.342			
cusl_3					0.313			
like_1						0.406		
like_2						0.303		
like_3						0.290		
perf_1							0.346	
perf_2							0.136	
perf_3							0.133	
perf_4							0.244	
perf_5							0.142	
qual_1								0.162
qual_2								0.031
qual_3								0.080
qual_4								-0.003
qual_5								0.125
qual_6								0.290
qual_7								0.165
qual_8								0.150

the rescaling is extremely limited. Therefore, we retain this indicator and continue the analysis. If the (negative) weight had been larger (e.g., −0.100), we would have had to delete this indicator for applying the IPMA. However, such a step has to be in line with measurement theory in that the remaining indicators still cover the construct domain sufficiently (see Hair, Hult, Ringle, & Sarstedt, 2022, Chapter 5).

Next, click on the **Edit** symbol in the toolbar to close the result report and to return to the SmartPLS **Modeling window**. We now run the IPMA by clicking on **Calculate → Importance-performance map analysis (IPMA)** in the toolbar. Alternatively, we can left-click on the **Calculate** wheel symbol in the toolbar and select the corresponding option in the combo box that opens. In the following dialog box (Exhibit 4.27), specify

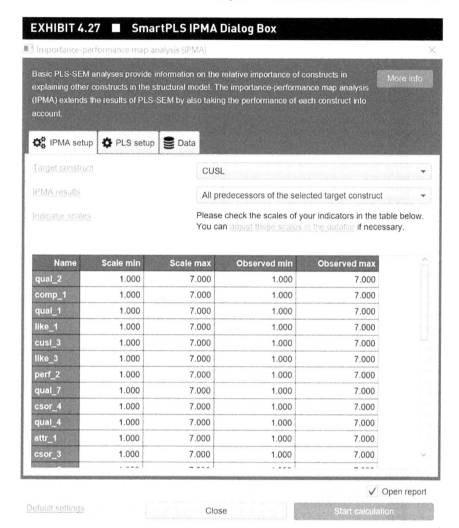

EXHIBIT 4.27 ■ SmartPLS IPMA Dialog Box

the **Target construct** *CUSL* and choose the **All predecessors of the selected target construct** option. In this dialog, we can double-check if the **Scale min** or **Scale max** has values of 1 and 7 for all indicators (i.e., this was the scale used in the survey). If this is not the case, the IPMA rescaling would be incorrect and we need to click on the **adjust these scales in the datafile** hyperlink. Ensure to select the **Unstandardized** option for **Type of results** in the PLS setup tab. Otherwise, use the default settings for all the other options and make sure to tick the box next to **Open report**. Finally, click on **Start calculation** (SmartPLS now automatically computes the performance and importance values (Steps 2 and 3) and creates the importance-performance map (Step 4).

After completing the computations, the **Results report** opens and PLS path model appears with the IPMA results (Exhibit 4.28). It is important to note that these results differ from the results obtained for the corporate reputation model

EXHIBIT 4.28 ■ **IPMA Results**

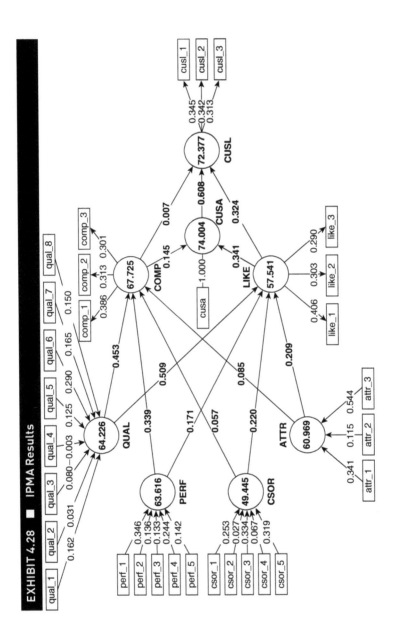

EXHIBIT 4.29 ■ **Importance-Performance Map (Construct Level) of Target Construct *CUSL***

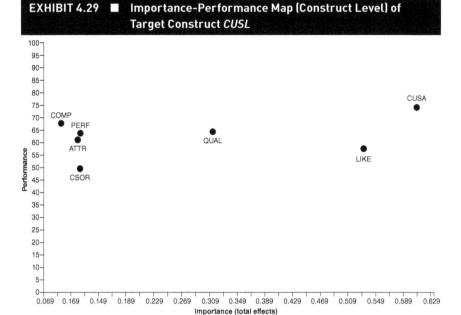

estimation shown in the other chapters of the book (e.g., Chapter 1). Instead of showing the R^2 values of the endogenous constructs in the PLS path model, the IPMA results show each construct's performance values. Furthermore, the IPMA results do not display the standardized indicator loadings or weights, but the unstandardized and rescaled indicator weights of the measurement models, regardless of whether they are formative or reflective.

In the **Results report**, click on **Quality criteria → Importance-performance map [CUSL] (constructs)** to show the importance-performance map as displayed in Exhibit 4.29. We can also request the unstandardized total effects and performance values in table format by clicking on **Construct total effects for [CUSL]** and **Construct performances for [CUSL]**, respectively. The table format proves useful in case we want to redraw the importance-performance map in a spreadsheet software such as Microsoft Excel. SmartPLS also allows for customizing the importance-performance map. To do so, click on **customize this chart** right below the map (Exhibit 4.29).

Not surprisingly, we find that two of the three direct predecessors, *CUSA* and *LIKE*, have a particularly high importance for *CUSL* (Exhibit 4.29). More importantly, *QUAL* has the greatest importance for *CUSL* compared to the other three exogenous constructs *ATTR, CSOR,* and *PERF*. Managerial or policy actions should, therefore, prioritize improving *CUSA* and *LIKE*, which can be achieved by focusing on the exogenous construct *QUAL*. Addressing *QUAL* is particularly promising because this construct has a formative measurement model, whose indicator weights indicate which aspect is most important for improving *QUAL*.

To gain more specific information on how to increase the performance of the constructs, particularly that of *QUAL*, the following analyses focus on the indicator level (Step 5). Note again that an IPMA on the indicator level is possible regardless of the predecessor constructs' measurement model specifications. However, an indicator-related analysis is particularly useful in formative measurement model settings. When going to **Quality criteria** → **Importance-performance map [CUSL] (indicators)** → **Importance-performance map** in the results report, SmartPLS shows the indicators' importance-performance map as displayed in Exhibit 4.30. However, this display of all indicators does not offer helpful guidance for identifying features that should be addressed in order to improve customer loyalty. For this reason, we focus on the formatively measured constructs in the PLS path model, because they represent relatively concrete features that can be addressed for performance improvements. Among these constructs, *QUAL* clearly has the greatest importance, while showing an average performance. Hence, we focus on the IPMA of *QUAL's* indicators and click on all indicators other than those of *QUAL* in the caption of the importance-performance map. By clicking on an indicator in the caption, it will disappear in the importance-performance map—clicking on the indicator again will make it appear again. Exhibit 4.31 shows the reduced importance-performance map with *QUAL's* indicators only.

Alternatively, under **Quality criteria** → **Importance-performance map [CUSL] (indicators)** in the results report, SmartPLS shows the indicators' unstandardized total effects in the **Indicator total effects for [CUSL]** tab and the performance values in the **Indicator performances for [CUSL]** tab. Exhibit 4.32 shows these outputs for *QUAL's*

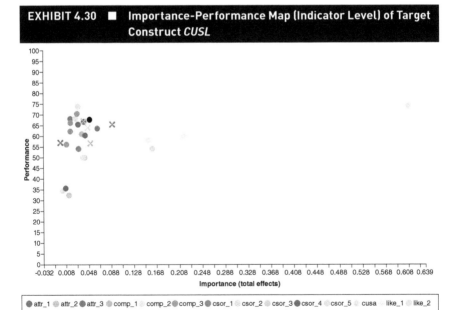

EXHIBIT 4.30 ■ **Importance-Performance Map (Indicator Level) of Target Construct *CUSL***

EXHIBIT 4.31 ■ **Importance-Performance Map (*QUAL* Indicator Level)**

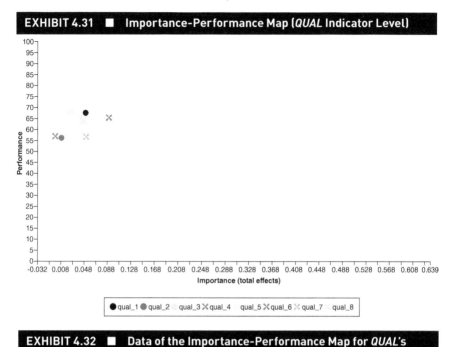

EXHIBIT 4.32 ■ **Data of the Importance-Performance Map for *QUAL*'s Indicators**

Indicators of QUAL	Importance	Performance
qual_1	0.051	67.539
qual_2	0.010	56.202
qual_3	0.025	68.023
qual_4	−0.001	56.880
qual_5	0.039	66.860
qual_6	0.091	65.407
qual_7	0.052	56.638
qual_8	0.047	63.953
Mean Value	0.039	62.688

indicators, which enable producing the importance-performance map as shown in Exhibit 4.33. The IPMA reveals that the indicator *qual_6* (i.e., [The company] is a reliable partner for customers), has a particularly high importance for improving customer loyalty. Similarly, *qual_7* (i.e., [The company] is a trustworthy company) has a relatively high importance but also shows a particularly low performance. These results suggest that companies should especially stress their reliability and trustworthiness when aiming to improve their customers' loyalty.

EXHIBIT 4.33 ■ Importance-Performance Map for *QUAL*'s Indicators

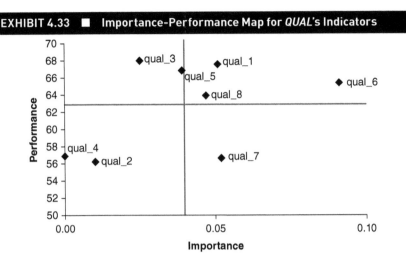

The results of the IPMA provide the empirical foundations for a better discussion on how to improve the performance of a key target construct. In particular, the IPMA identifies specific indicators with high importance but relatively low performance, which then become the focal areas for managing performance improvements activities. Subsequent qualitative and quantitative analyses might include additional important factors, such as costs and time, and, as a result, could entail a revised prioritization of activities. For example, an area with high priority for performance improvements, as revealed by the IPMA, might drop into a lower priority rank, if its performance improvements are very costly and require much time. As a result of these analyses, we obtain a comprehensive picture of effective actions for the overall performance improvement opportunities regarding our key construct of interest (e.g., customer loyalty).

SUMMARY

- **Understand why and how to run a necessary condition analysis (NCA) in PLS-SEM.** An independent construct can be sufficient to produce a change in a dependent variable (i.e., sufficiency logic). However, it might not be necessary for producing the change and its absence could be compensated for by other independent constructs. Such a necessity logic implies that a certain level of an outcome can only be achieved if the necessary construct is in place and at a certain level. The NCA enables the identification of necessary conditions by using construct scores from a previous PLS-SEM analysis. Rather than analyzing average relationships between dependent and independent constructs, as is done in a standard PLS-SEM analysis, the NCA reveals particular necessary conditions and the level of independent constructs required to achieve a certain outcome level.

- **Comprehend how to use SmartPLS to run an NCA in PLS-SEM.** The PLS-SEM-based NCA should use unstandardized construct scores as input and the analysis should consider the empirical scale. However, if all indicators are on the same scale (e.g., 1 to 7) and the resulting construct scores do not achieve these expected minimum and maximum values, the researcher should use the theoretical scale (e.g., 1 to 7). To identify whether or not necessity conditions have been met, researchers should rely on the necessity effect size d, which should be 0.1 or higher and statistically significant. To interpret results, the bottleneck tables and ceiling line charts are considered with the aim of disclosing values in the independent construct(s) that are required to achieve a certain outcome. Additional analyses identify observations that do not meet minimum threshold values in one or more constructs.

- **Appreciate the importance-performance map analysis (IPMA).** The purpose of the IPMA is to extend the standard PLS-SEM results reporting of path coefficient estimates by adding a dimension that considers the average values of the construct scores. More precisely, the IPMA contrasts the total effects, representing the predecessor constructs' importance in shaping a certain target construct, with their average construct scores indicating their performance. The goal is to identify predecessors that have a relatively high importance for the target construct (i.e., those that have a great total effect) but also have a relatively low performance (i.e., low average construct scores).

- **Comprehend how to conduct the IPMA using the SmartPLS software.** Before running an IPMA in SmartPLS, researchers need to check several requirements, which relate to the indicator scaling and the indicator weights as obtained from a standard PLS-SEM analysis. The IPMA requires selecting a key target construct and specifying the minimum and maximum values of the indicator scales. SmartPLS produces importance-performance maps on the construct level as well as on the indicator level.

REVIEW QUESTIONS

1. Why do we need to distinguish between sufficiency logic and necessity logic?

2. What is an NCA? What are the NCA's methodological strengths and weaknesses?

3. Why should researchers jointly use PLS-SEM and NCA?

4. What is the purpose of the IPMA?

5. When looking at the importance-performance map, what does a ceteris paribus interpretation of the results mean?

CRITICAL THINKING QUESTIONS

1. What is the CE-FDH line, and why is it important to consider this line when following sufficiency versus necessity logics?

2. What is the CR-FDH line, and why is it important to consider this line when following sufficiency versus necessity logics?

3. Under which circumstances would you assume that an independent construct is merely sufficient, not necessary?

4. How can you systematically draw practical implications from the IPMA results? When and how would you include the indicator layer?

5. Name an additional (third) interesting dimension of an IPMA that you would consider. Why and how can these additional considerations change your previous prioritization of activities?

KEY TERMS

Bottleneck table

Ceiling envelopment free disposal hull (CE-FDH) line

Ceiling regression free disposal hull (CR-FDH) line

Direct effect

Equidistant scale

Forced-choice scale

Impact-performance map

Importance

Importance-performance map analysis (IPMA)

Importance-performance matrix

Indirect effect

Metric scale

Necessary condition analysis (NCA)

Necessity effect size d

Necessity logic

Nominal scale

Ordinal scale

Performance

Priority map analysis

Rescaling

Sufficiency logic

Total effect

SUGGESTED READINGS

Bokrantz, J., & Dul, J. (2023). Building and testing necessity theories in supply chain management. *Journal of Supply Chain Management, 59*(1), 48–65.

Dul, J. (2016). Necessary condition analysis (NCA): Logic and methodology of "necessary but not sufficient" causality. *Organizational Research Methods, 19*, 10–52.

Dul, J. (2020). *Conducting Necessary Condition Analysis.* Sage.

Hauff, S., Guerci, M., Dul, J., & van Rhee, H. (2021). Exploring necessary conditions in HRM research: Fundamental issues and methodological implications. *Human Resource Management Journal, 31*, 18–36.

Höck, C., Ringle, C. M., & Sarstedt, M. (2010). Management of multi-purpose stadiums: Importance and performance measurement of service interfaces. *International Journal of Services Technology and Management*, *14*(2–3), 188–207.

Mikulic, J., Prebežac, D., & Dabic, M. (2016). Importance-performance analysis: Common misuse of a popular technique. *International Journal of Market Research*, *58*(6), 775–778.

Richter, N. F., & Hauff, S. (2022). Necessary conditions in international business research: Advancing the field with a new perspective on causality and data analysis. *Journal of World Business*, *57*, 101310.

Richter, N. F., Hauff, S., Ringle, C. M., & Gudergan, S. P. (2022). The use of partial least squares structural equation modeling and complementary methods in international management research. *Management International Review*, *62*, 449–470.

Richter, N. F., Hauff, S., Ringle, C. M., Sarstedt, M., Kolev, A., & Schubring, S. (2023). How to apply necessary condition analysis in PLS-SEM. In J. F. Hair, R. Noonan, & H. Latan (Eds.), Partial least squares structural equation modeling: *Basic concepts, methodological issues and applications* (2nd ed.), forthcoming. Springer.

Richter, N. F., Schubring, S., Hauff, S., Ringle, C. M., & Sarstedt, M. (2020). When predictors of outcomes are necessary: Guidelines for the combined use of PLS-SEM and NCA. *Industrial Management & Data Systems*, *120*, 2243–2267.

Ringle, C. M., & Sarstedt, M. (2016). Gain more insight from your PLS-SEM results: The importance-performance map analysis. *Industrial Management & Data Systems*, *116*(9), 1865–1886.

Schloderer, M. P., Sarstedt, M., & Ringle, C. M. (2014). The relevance of reputation in the nonprofit sector: The moderating effect of socio-demographic characteristics. *International Journal of Nonprofit and Voluntary Sector Marketing*, *19*(2), 110–126.

Sukhov, A., Olsson, L. E., & Friman, M. (2022). Necessary and sufficient conditions for attractive public transport: Combined use of PLS-SEM and NCA. *Transportation Research Part A: Policy and Practice*, *158*, 239–250.

van der Valk, W., Sumo, R., Dul, J., & Schroeder, R. G. (2016). When are contracts and trust necessary for innovation in buyer-supplier relationships? A necessary condition analysis. *Journal of Purchasing and Supply Management*, *22*(4), 266–277.

MODELING OBSERVED HETEROGENEITY

CHAPTER PREVIEW

Relationships in PLS path models imply that exogenous constructs explain endogenous constructs without any systematic influences of other variables. In many instances, however, this assumption does not hold. For example, respondents are likely to be heterogeneous in their perceptions and evaluations of constructs, yielding significant differences in, for example, path coefficients across two or more groups of respondents (e.g., retail vs. wholesale customers). Recognizing that heterogeneous data structures are often present, researchers are increasingly interested in identifying and understanding such differences. In fact, failure to consider heterogeneity can be a threat to the validity of PLS-SEM results (Becker, Rai, Ringle, & Völckner, 2013; Sarstedt & Ringle, 2010).

As a solution, the following two chapters introduce different concepts that enable researchers to examine and model heterogeneous data. This chapter first provides an overview of observed and unobserved heterogeneity, including describing how disregarding heterogeneous data structures can produce biased results. Next, we discuss measurement invariance, which is a primary concern and must be assessed before comparing groups of data, regardless of the source of heterogeneity. By establishing measurement invariance, researchers can be confident that group differences in model estimates result from neither the distinctive content and/or meanings of the constructs across groups nor the measurement scale. To assess measurement invariance in a PLS-SEM context, researchers can use the measurement invariance of composite models (MICOM) procedure.

Next, we introduce different types of multigroup analyses used to compare parameters (usually path coefficients) between two or more groups of data. A PLS-SEM multigroup analysis is typically applied when researchers want to explore differences that can be traced back to observable characteristics such as gender or country of origin. In this situation, researchers assume that there exists a categorical moderator variable (e.g., customer type) which influences the relationships in the PLS path model. Yet, while categorical data allow easy specification of groups, other data can also provide a basis for specifying groups. The aim of multigroup analysis, therefore, is to identify the effect of this categorical variable. The chapter closes with an illustration of measurement model invariance assessment using the MICOM procedure and a multigroup analysis.

OBSERVED AND UNOBSERVED HETEROGENEITY

Applications of PLS-SEM usually analyze the full set of data, implicitly assuming the data stem from a homogeneous population. This assumption of relatively homogeneous data characteristics is often unrealistic. Individuals (e.g., in their attitudes), companies (e.g., in their structure), or environments (e.g., in their dynamics) are frequently different, and pooling data across observations is likely to produce misleading results. Failure to consider heterogeneity can be a threat to the validity of PLS-SEM results since it can lead to incorrect conclusions (Sarstedt & Ringle, 2010). Becker, Rai, Ringle, and Völckner (2013) address this problem in detail and provide examples of situations with significantly different positive and negative path coefficients, which show a nonsignificant value close to 0 on the aggregate data level. Concluding that no relationship exists between the constructs would be invalid and misleading.

The model shown in Exhibit 5.1, in which customer satisfaction with a product (Y_3) depends on satisfaction with the quality (Y_1) and satisfaction with the price (Y_2), illustrates the problems stemming from failure to treat heterogeneity in the context of PLS-SEM. Suppose there are two segments of similar sizes. Segment 1 comprises retail customers, whereas segment 2 consists of wholesale customers. Both segments differ regarding their price consciousness as indicated by the different segment-specific path coefficients. More precisely, the effect of perceived quality (Y_1) on customer satisfaction (Y_3) is much stronger among retail customers $(p_{13}^{(1)} = 0.50$, i.e., group 1; the superscript in parentheses indicates the group) than among wholesale customers $(p_{13}^{(2)} = 0.10$, i.e., group 2). In contrast, perceived price (Y_2) has a somewhat stronger influence on customer satisfaction (Y_3) among wholesale customers $(p_{23}^{(2)} = 0.35)$ than among retail customers $(p_{23}^{(1)} = 0.25)$. In this example, heterogeneity reflects the results of one segment (wholesale customers) being more price sensitive but less quality sensitive, whereas it is the opposite for the other segment (retail customers). From a technical perspective, there is a categorical moderator variable *customer type* that splits the dataset into two customer segments and thus allows estimating the model for two separate data groups, as indicated in Exhibit 5.1. Importantly, if we fail to recognize the heterogeneity between the groups and analyze the model using the full set of data,

EXHIBIT 5.1 ■ Heterogeneity in PLS Path Models

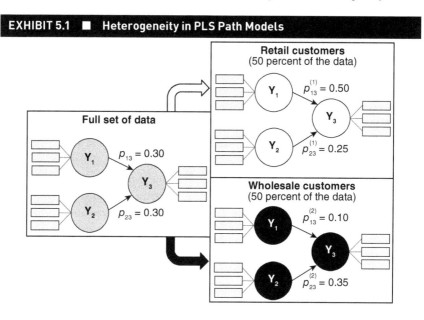

the path coefficient estimates offer an incomplete picture of the model relationships. That is, both estimates would equal approximately 0.30 when using the full set of data, thus leading the researcher to conclude that price and quality are equally important for retail and wholesale customers, although they are not. Similarly, segment-related differences in model estimates can cancel each other out, yielding nonsignificant effects when analyzing the data on an aggregate level (for an example, see Sarstedt, Schwaiger, & Ringle, 2009). Consequently, it is important to identify, assess, and, if present, treat heterogeneity in the data.

Heterogeneity in data can be observed or unobserved. **Observed heterogeneity** occurs when differences between two or more groups of data relate to observable characteristics, such as gender, age, or country of origin. Researchers can use these observable characteristics to partition the data into separate groups of observations and carry out group-specific PLS-SEM analyses with regard to the customer type, as illustrated in Exhibit 5.1. On the contrary, **unobserved heterogeneity** implies that differences between two or more groups of data do not emerge a priori from a specific observable characteristic or combinations of several characteristics. Instead, they become apparent in differences in structural path coefficients.

To account for unobserved heterogeneity with regard to endogenous but also exogenous variables, researchers have routinely used **cluster analysis** techniques, such as **k-means clustering** (Hair, Black, Babin, & Anderson, 2019, Chapter 4; Sarstedt & Mooi, 2019, Chapter 9) on the indicator data, or construct scores derived from a previous analysis of the entire dataset. The partition produced by this analysis is then used as input for group-specific PLS-SEM estimations. While easy to apply, such an approach is conceptually flawed because traditional clustering techniques ignore the path model relationships that researchers specified prior to the analysis. However, exactly these relationships are likely responsible for some of the group differences. Therefore, it is not surprising that prior research has shown traditional clustering approaches to perform poorly in identifying group differences in PLS-SEM (e.g., Sarstedt & Ringle, 2010).

Recognizing the limitations of sequential approaches, methodological research in PLS-SEM has proposed several specific methods to identify and treat unobserved heterogeneity, commonly referred to as **latent class techniques**. These techniques have a long tradition in CB-SEM research (e.g., Jedidi, Jagpal, & DeSarbo, 1997; Masyn, 2013; Muthén, 1989) and have proven very useful in identifying unobserved heterogeneity and partitioning the data accordingly. The resulting partition can then be analyzed for significant differences using multigroup analysis approaches. Alternatively, latent class techniques might ascertain that unobserved heterogeneity does not influence the results, supporting an analysis of a single model based on the aggregate-level data.

The different techniques for identifying unobserved heterogeneity in PLS-SEM include genetic algorithms (Ringle, Sarstedt, & Schlittgen, 2014; Ringle, Sarstedt, Schlittgen, & Taylor, 2013), weighted least squares (Schlittgen, Ringle, Sarstedt, & Becker, 2016), and several other approaches (e.g., Becker, Rai, Ringle, & Völckner, 2013; Esposito Vinzi, Trinchera, Squillacciotti, & Tenenhaus, 2008)—see Hair, Sarstedt, Matthews, and Ringle (2016) for a review. In Chapter 6, we introduce two latent class procedures, **finite mixture partial least squares (FIMIX-PLS)**, and **prediction-oriented segmentation in PLS-SEM (PLS-POS)**, whose joint use facilitates reliably identifying and treating unobserved heterogeneity in PLS path models.

TESTING FOR MEASUREMENT MODEL INVARIANCE

A primary concern before comparing group-specific parameter estimates for significant differences using a multigroup analysis—regardless of whether the groups have been formed based on observable characteristics or stem from a latent class analysis—is ensuring **measurement invariance**, also referred to as **measurement equivalence**. By establishing measurement invariance, researchers can be confident that group differences in model estimates do not result from the distinctive content and/or meanings of the constructs across groups. Variations in the structural relationships between constructs could emerge from different meanings the groups' respondents attribute to the phenomena being measured, rather than the true differences in the structural relationships. Reasons for such differences might stem from, for example, (1) respondents who embrace different cultural values and, consequently, interpret a given measure in a conceptually different manner; (2) gender, ethnic, or other individual differences that entail responding to instruments in systematically different ways; and (3) respondents who use the available options on a scale differently (e.g., tendency to choose or not to choose the extremes). Hult, Griffith, Finnegan, Gonzalez-Padron, Harmancioglu,

Huang, Talay, & Cavusgil, (2008, p. 1028) describe these concerns and conclude that "failure to establish data equivalence is a potential source of measurement error" (i.e., discrepancies between what is intended to be measured and what is actually measured). When measurement invariance is not present, it can render statistical tests less meaningful and provide misleading results. In short, when measurement invariance is not demonstrated, any conclusions about model relationships are questionable. Hence, multigroup comparisons require establishing measurement invariance to ensure the validity of outcomes and conclusions.

Researchers have suggested a variety of methods to assess measurement invariance for CB-SEM. Multigroup confirmatory factor analysis based on the guidelines of Steenkamp and Baumgartner (1998) and Vandenberg and Lance (2000) is by far the most common approach to invariance assessment. However, the well-established measurement invariance techniques used to assess CB-SEM's **common factor models** and related extensions to formative measurement models (Diamantopoulos & Papadopoulos, 2010) cannot be readily transferred to PLS-SEM's composite models. For this reason, Henseler, Ringle, and Sarstedt (2016) developed the **measurement invariance of composite models (MICOM) procedure**. The MICOM procedure builds on the construct scores, which result from linear combinations (i.e., composites) of indicators and the indicator weights as estimated by the PLS-SEM algorithm. Therefore, in the description of the MICOM procedure, we talk about composites when referring to the entities (scores) the PLS-SEM algorithm uses to represent the constructs as specified by the researcher.

The MICOM procedure involves three steps: (1) **configural invariance**, (2) **compositional invariance**, and (3) **equality of composite mean values and variances**. The three steps are hierarchically interrelated, as displayed in Exhibit 5.2. This means configural invariance is a precondition for compositional invariance, which again is a precondition for meaningfully assessing the equality of composite mean values and variances.

If configural invariance (Step 1) and compositional invariance (Step 2) are established, **partial measurement invariance** is confirmed. When partial measurement invariance is confirmed for all constructs in the PLS path model, researchers can compare the path coefficients by means of a multigroup analysis. If partial measurement invariance is established and, additionally, the composites have equal mean values (Step 3a) and variances (Step 3b) across the groups, **full measurement invariance** is confirmed. From a measurement model perspective, researchers can then pool the data of the different groups, if they come from separate groups initially, and benefit from the increase in statistical power. However, full measurement invariance does not imply that there are no differences in the structural model across groups (i.e., the path coefficients). The latter need to be tested by means of a multigroup analysis. Only if the multigroup analysis indicates that the structural models are also invariant (i.e., there are no significant differences in the path coefficients), researchers can pool the data and exclusively focus on the aggregate-level analysis. In the following sections, we

EXHIBIT 5.2 ■ The MICOM Procedure

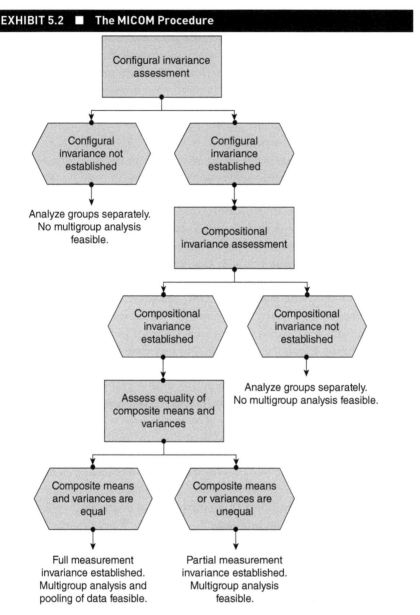

discuss each step in greater detail (for an application of the MICOM on data from five countries, see Schlägel & Sarstedt, 2016).

Step 1: Configural Invariance

Step 1 addresses the establishment of configural invariance to ensure that each construct in the PLS path model has been specified equally for all the groups. Configural variance exists when constructs are equally parameterized and estimated across groups.

An initial qualitative assessment of the constructs' specification across all the groups is required to ensure the following three requirements have been met:

1. *Identical indicators per measurement model.* Each measurement model must employ the same indicators and scales across the groups. Checking whether exactly the same indicators apply to all the groups seems rather simple. However, when conducting surveys using different languages, the application of good empirical research practices (e.g., translation and back-translation) is of the utmost importance to establish the indicators' equivalence. In this context, an assessment of face and/or expert validity can help verify whether the researcher(s) used the same set of indicators across the groups.

2. *Identical data treatment.* The indicators' data treatment must be identical across all the groups, which includes the coding (e.g., dummy coding), reverse coding, and other forms of recoding as well as data handling (e.g., standardization or missing value treatment). Outliers should be detected and treated similarly.

3. *Identical algorithm settings or optimization criteria.* Variance-based model estimation methods such as PLS-SEM consist of many variants with different target functions and algorithm settings (e.g., choice of initial outer weights and the inner model weighting scheme; Hair, Ringle, & Sarstedt, 2011; Henseler, Ringle, & Sinkovics, 2009). Researchers must ensure that differences in the group-specific model estimations do not result from dissimilar algorithm settings.

Configural invariance is a necessary but not sufficient condition for drawing valid conclusions from multigroup analyses. Researchers must also ensure that differences in the path coefficients do not result from differences in the way a construct is formed across the groups. The next step, compositional invariance, focuses on this aspect.

Step 2: Compositional Invariance

Compositional invariance exists when the composite scores are the same across the groups, despite possible differences in the group-specific weights used to compute the scores. Step 2 of the MICOM procedure applies a statistical test to assess whether the composite scores differ significantly across the groups. For this purpose, the MICOM procedure examines c, which is the correlation between the composite scores $Y^{(1)}$ and $Y^{(2)}$:

$$c = \text{cor}\left(Y^{(1)}, Y^{(2)}\right)$$

Compositional invariance requires that c equals 1. Technically, the procedure tests the null hypothesis that c is 1. In order to establish compositional invariance, we must not reject this null hypothesis, for instance, at a significance level of 5%. That is, if the test yields a p value larger than 0.05, we can assume compositional invariance. On the contrary, if we reject the null hypothesis (i.e., the test yields a p value smaller than 0.05), we cannot establish compositional invariance for the specific construct under consideration. Note that this test does not work for single-item constructs, since their (single) outer relationship is always 1. Hence, the construct scores $Y^{(1)}$ and $Y^{(2)}$ of a single-item construct are identical, resulting in a c value of 1. Note that in exploratory research settings or when the group-specific sample sizes are small, assuming a significance level of 1% is more reasonable. Assuming a lower significance level implies that compositional invariance is more likely to be established.

For hypothesis testing, the MICOM procedure draws on the concept of **permutation** (Fisher, 1935). Similar to bootstrapping, permutation tests generate a reference distribution from the actual data. However, instead of sampling observations from the original dataset with replacement—as it is the case in bootstrapping—permutation tests randomly exchange observations between the groups multiple times. By comparing the parameter estimates of the permuted datasets with those of the original one, researchers can assess whether observed effects are systematic. For example, assume that the estimated correlation between the sets of construct scores c is close to 1 in the original comparison of the two datasets. If the correlation fluctuates strongly when permuting the observations between the two groups, this suggests that there is a systematic reason as to why the original correlation is close to 1—the original data partition. Permutation testing has routinely been used in a multitude of contexts since it provides an efficient approach to nonparametric testing, also when the sample size is small (e.g., Ernst, 2004; Good, 2000).

The compositional invariance assessment in the MICOM follows a five-step approach, which Exhibit 5.3 illustrates:

1. Group-specific estimation of the PLS path models to obtain composite scores of group 1 with $n^{(1)}$ observations and group 2 with $n^{(2)}$ observations. Note that in the example of two groups the sum of $n^{(1)}$ and $n^{(2)}$ equals n, which is the total number of observations in the full set of data. Also note that the example uses only four observations per group for illustrative purposes.

2. Computation of the correlations c between the composite scores of groups 1 and 2.

3. Random permutation of the data, meaning that the $n^{(1)}$ observations are randomly assigned to the groups. In formal terms, observations are randomly drawn without replacement from the aggregate dataset and are assigned to group 1. The remaining $n^{(2)}$ observations are assigned to group 2. Researchers should use a minimum of 1,000 permutations.

4. For each permutation run u, compute each composite's correlation c_u between the scores of groups 1 and 2.

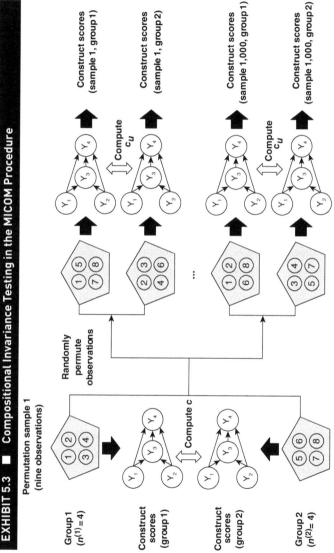

EXHIBIT 5.3 ■ Compositional Invariance Testing in the MICOM Procedure

5. For hypothesis testing, the procedure sorts the permutations' correlation results c_u in descending order. The cutoff value is the 95% point (or, in our example, the 950th of 1,000 permutations in the sorted list) and the corresponding correlation value. If c is smaller than that value, it falls out of the 95% permutation-based **confidence interval** and is significantly different from 1 (at the $p < 0.05$ level), which entails a rejection of the null hypothesis that c equals 1. Consequently, the composite scores are significantly different, and compositional invariance has not been established. In the opposite situation, if c does not fall into the 5% extreme tail, but into the 95% confidence interval, the composite scores are not significantly different from one another, which substantiates compositional invariance. Based on the permutation results, it is also possible to return the p value of c. In this analysis, the p value represents the percentage of the permutations' correlation c_u results that have a lower value than c. A p value above 0.05 indicates that c is not significantly different from 1, which means that compositional invariance has been established.

When configural invariance exists and compositional invariance is established for all constructs in the PLS path model, there is partial measurement invariance. In this case, researchers can compare the path coefficients by means of a multigroup analysis. If partial measurement invariance cannot be established for a construct, then group-specific comparisons using a multigroup analysis on any relationship involving this construct are not feasible. In this case, researchers should refrain from running a multigroup analysis altogether and analyze each group separately without attempting to compare the results for the groups. Alternatively, one can eliminate the constructs that did not achieve compositional invariance, provided theory supports this step. Importantly, when making any changes to the model, the MICOM procedure needs to be rerun.

Step 3: Equality of Composite Mean Values and Variances

If the results of Step 2 support measurement invariance, the assessment should continue with the equality assessment of the composites' mean values and variances.

This final MICOM step requires estimating the PLS path model using the pooled (i.e., aggregate) set of data. The procedure then partitions the dataset according to the initial grouping and computes the composites' means and variances for each group. Note that these means and variances are not 0 and 1 as they would be if we used the aggregate dataset for computation. Instead, we take the construct score estimates from the aggregate analysis and split up the values ex post according to the grouping variable. We then examine whether the mean values and variances between the composite scores of group 1 and the composite scores of group 2 differ regarding their means and variances. For the analysis of the mean values' equivalence, the null hypothesis is

$$H_0 : \overline{Y}^{(1)}_{pooled} - \overline{Y}^{(2)}_{pooled} = 0.$$

Here, the index "pooled" indicates that the composite scores of a certain group are calculated by the analysis of the pooled data.

The variance equivalence analysis requires computing the logarithm of the composite scores' variance ratio of both groups. If the logarithm of this ratio is statistically not different from 0, we conclude that the variances are equal across groups. The corresponding null hypothesis is

$$\text{H}_0 : \log\left(\text{var}\left(Y^{(1)}_{\text{pooled}}\right) / \text{var}\left(Y^{(2)}_{\text{pooled}}\right)\right) = \log\left(\text{var}\left(Y^{(1)}_{\text{pooled}}\right)\right) - \log\left(\text{var}\left(Y^{(2)}_{\text{pooled}}\right)\right) = 0.$$

The testing of these two hypotheses follows a similar approach to the one in Step 2. MICOM permutes (i.e., rearranges) the observations' group membership many times and generates the empirical distribution of the differences in mean values and logarithms of variances. Full measurement invariance is established when there are no significant differences in mean values and (logarithms of) variances across the groups. This is the case when the mean values and variances as obtained by the original model estimation (i.e., without permutation) fall into the permutation-based confidence intervals of the differences in mean values and (logarithms of) variances. Analogous to Step 2, p values larger than 0.05 (assuming a 5% significance level) indicate that there are no differences in means and variances. In contrast, if any of the original differences fall outside the confidence interval (i.e., the difference is significant), we must acknowledge that full measurement invariance has not been established.

In summary, comparing group-specific model relationships for significant differences using a multigroup analysis requires establishing configural (Step 1) and compositional (Step 2) invariance. If these two steps support measurement invariance, partial measurement invariance is established. As the focus is usually on the structural model relationships, we can then run a multigroup analysis on the path coefficients using one of the approaches described in the following sections. However, an exception in this regard is the comparison of interaction terms, created in the course of a moderator analysis (see Hair, Hult, Ringle, & Sarstedt, 2022, Chapter 7). The measurement model of an interaction term represents an auxiliary measurement that incorporates the interrelationships between the moderator and the exogenous construct in the path model. This characteristic, however, renders any measurement model assessment of the interaction term and related comparisons across different groups in terms of invariance assessment meaningless. Importantly, if partial measurement invariance cannot be established for a construct, group-specific comparisons using a multigroup analysis on any relationship involving this construct are not feasible, including moderated or mediated ones.

To reiterate, if partial measurement invariance is not established, researchers should analyze the groups separately without drawing any conclusions about group-related effects. Alternatively, they may consider changing the model by eliminating the constructs that did not achieve compositional invariance—provided that structural theory supports such a step. Finally, if partial measurement invariance is confirmed and the composites have equal mean values and variances across the groups, full measurement invariance is confirmed, which additionally supports the analysis of the pooled data, when considered applicable.

MULTIGROUP ANALYSIS

Path coefficients generated from different samples are almost always numerically different, but the question is whether the differences are statistically significant. **Multigroup analysis (MGA)** helps answer this question (Chin & Dibbern, 2010; Matthews, 2017; Sarstedt, Henseler, & Ringle, 2011). Technically, an MGA tests the null hypotheses H_0 that the path coefficients between two groups (e.g., $p^{(1)}$ in group 1 and $p^{(2)}$ in group 2) are not significantly different (e.g., $p^{(1)} = p^{(2)}$), which amounts to the same as saying that the absolute difference between the path coefficients is 0 (i.e., $H_0 : \left| p^{(1)} - p^{(2)} \right| = 0$). The corresponding alternative hypothesis H_1 is that the path coefficients are different (i.e., $H_1 : p^{(1)} \neq p^{(2)}$ or, put differently, $H_1 : \left| p^{(1)} - p^{(2)} \right| \neq 0$.

Research has proposed various approaches for comparing a single path coefficient across two (or more) groups, which can be differentiated into parametric and nonparametric approaches, depending on whether they rely on any distributional assumptions (Matthews, 2017; Sarstedt, Henseler, & Ringle, 2011). Recent research has compared the performance of different approaches using simulated data (Klesel, Schubert, Niehaves, & Henseler, 2022), identifying approaches that perform satisfactorily in situations commonly encountered in empirical research. These include the parametric MGA, and two nonparametric approaches—the bootstrap MGA and the permutation MGA (Exhibit 5.4).

Research has also proposed a test to evaluate whether the complete model is different across two (or more) groups (Klesel, Schuberth, Henseler, & Niehaves, 2019)—see Exhibit 5.4. The **nonparametric distance-based test** applies the permutation procedure to compare the average (squared Euclidean or geodesic) distance of the model-implied indicator correlation matrix across the groups. Klesel, Schubert, Niehaves, and Henseler's (2022) follow-up study provides support for the efficacy of the test when the

EXHIBIT 5.4 ■ Multigroup Analysis Approaches in PLS-SEM

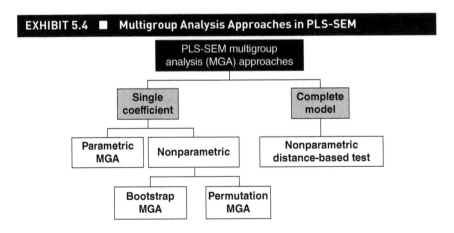

objective is to compare the overall model across groups. However, the nonparametric distance-based test only indicates that group difference can exist in the overall model comparison, but it does not provide clear parameter information on whether there are specific differences between path coefficients in group comparisons (Cheah, Amaro, & Roldán, 2023). Therefore, researchers can use the test to demonstrate in general that group differences exist, but they cannot show them specifically. Since this situation is rarely satisfactory in practical applications of PLS-SEM, the following discussion focuses on the case of comparing a single path coefficient across two (or more) groups.

Parametric MGA

The **parametric MGA** (Keil, Saarinen, Tan, Tuunainen, Wassenaar, & Wei, 2000) was the first approach applied in PLS-SEM studies and has been widely adopted because of its ease of implementation. This approach is a modified version of a standard two independent samples t test, which relies on standard errors derived from **bootstrapping**. Similar to the standard t test, the parametric approach has two versions (Sarstedt & Mooi, 2019), depending on whether population variances of the composite scores that are being used as input for the computation of the path coefficients can be assumed to be equal (homoscedastic) or unequal (heteroscedastic). To identify whether the variances are equal or unequal, researchers can rely on the results of Step 3 of the MICOM procedure. Specifically, if the original variance difference between two groups falls inside the percentile confidence interval or when the p value of this difference is higher than the prespecified significance level, equal variances should be assumed.

If the composite score variances are *equal*, the **parametric MGA (equal)** test statistic (i.e., the empirical t value) can be applied and is computed as follows:

$$t = \frac{p^{(1)} - p^{(2)}}{\sqrt{\frac{(n^{(1)} - 1)^2}{(n^{(1)} + n^{(2)} - 2)} \cdot se\left(p^{(1)}\right)^2 + \frac{(n^{(2)} - 1)^2}{(n^{(1)} + n^{(2)} - 2)} \cdot se\left(p^{(2)}\right)^2} \cdot \sqrt{\frac{1}{n^{(1)}} + \frac{1}{n^{(2)}}}}$$

In this formula, $p^{(1)}$ $\left(p^{(2)}\right)$ describes the path coefficient to be compared in group 1 (group 2), whereas $n^{(1)}$ $\left(n^{(2)}\right)$ stands for the number of observations in group 1 (group 2). Finally, $se\left(p^{(1)}\right)$ and $se\left(p^{(2)}\right)$ describe the standard errors of the parameter estimates of groups 1 and 2, respectively. These standard errors can be obtained via bootstrapping. To reject the null hypothesis of equal path coefficients, the empirical t value must be larger than the critical value of a t distribution with $n^{(1)} + n^{(2)} - 2$ degrees of freedom.

On the contrary, if the composite score variances are *unequal*, researchers should use a modified version of the above test statistic, referred to as **parametric MGA (unequal),** which relies on the **Welch-Satterthwaite t test** (Satterthwaite, 1946; Welch, 1947). Its test statistic takes the following form:

$$t = \frac{p^{(1)} - p^{(2)}}{\sqrt{\frac{(n^{(1)} - 1)}{n^{(1)}} \cdot se\left(p^{(1)}\right)^2 + \frac{(n^{(2)} - 1)}{n^{(2)}} \cdot se\left(p^{(2)}\right)^2}}$$

This test statistic is also asymptotically t distributed, but with the following degrees of freedom (df):

$$df = \left\| \frac{\left[\frac{(n^{(1)} - 1)}{n^{(1)}} \cdot se\left(p^{(1)}\right)^2 + \frac{(n^{(2)} - 1)}{n^{(2)}} \cdot se\left(p^{(2)}\right)^2 \right]^2}{\frac{(n^{(1)} - 1)}{n^{(1)}} \cdot se\left(p^{(1)}\right)^4 + \frac{(n^{(2)} - 1)}{n^{(2)}} \cdot se\left(p^{(2)}\right)^4} - 2 \right\|$$

Bootstrap MGA

Henseler, Ringle, & Sinkovics (2009) proposed the nonparametric **bootstrap MGA** that builds on bootstrapping results of each data group. For a specific relationship in the PLS path model, their approach compares each bootstrap estimate of one group with all other bootstrap estimates of the same parameter in the other group. By counting the number of occurrences in which the bootstrap estimate of group 1 is larger than those of group 2, the approach derives a p value for a one-tailed test. This approach can be described in formal terms as follows:

$$p\left(p^{(1)} \geq p^{(2)} | \beta^{(1)} < \beta^{(2)}\right) = 1 - \frac{1}{B^2} \sum_j \sum_j \Theta\left[\left(p_j^{(1)} + p^{(1)} - p^{-1}\right) - \left(p_j^{(2)} + p^{(2)} - p^{-2}\right)\right]$$

With regard to the population parameters $\beta^{(1)}$ and $\beta^{(2)}$ of each data group, this approach determines the conditional probability $p\left(p^{(1)} \geq p^{(2)} | \beta^{(1)} < \beta^{(2)}\right)$ that the $p^{(1)}$ estimation of a certain path $p^{(2)}$ coefficient in group 1 is larger than or equal to its estimated coefficient in group 2, whereby the superscript in parentheses marks the respective group. By using the bootstrap MGA, a researcher can examine whether this conditional probability is below a specified α-level before concluding that $\beta^{(1)}$ is greater than $\beta^{(2)}$. In the above equation, B is the number of bootstrap runs; $p_j^{(1)}$ ($j = 1, \ldots, B$) and $p_j^{(2)}$ ($j = 1, \ldots, B$) are the group-specific bootstrap results of the path coefficients; $\bar{p}^{(1)}$ and $\bar{p}^{(2)}$ denote the mean values of the group-specific bootstrap samples; and Θ (i.e., the Greek letter *theta*) stands for the unit step function, which has a value of 0 or 1 if its argument exceeds 0. Note that in accordance with Henseler, Ringle, & Sinkovics (2009), all group-specific bootstrap results need to be corrected by the difference between the original group-specific parameter estimate and the average over the group-specific bootstrap values.

To illustrate the working principle of the bootstrap MGA, consider a simple model with two constructs, estimated in two groups yielding the path coefficients $p^{(1)} = 0.336$ and $p^{(2)} = 0.501$. We like to test the hypothesis that the path coefficient is larger in group 1 compared to group 2. To test this hypothesis, we draw 10 **bootstrap samples** for $p^{(1)}$ and $p^{(2)}$ and contrast the estimates as shown in Exhibit 5.5. The first column represents the 10 bootstrap samples of the path coefficient in group 1, whereas the first row represents the 10 bootstrap samples in group 2. We now compare each bootstrap estimate of $p^{(1)}$ (e.g., the first bootstrap sample's estimate 0.357) with each bootstrap estimate of $p^{(2)}$ (i.e., 0.494, 0.423, ..., 0.538) and count the number of cases in which $p^{(1)} \geq p^{(2)}$, indicated by an X in Exhibit 5.5. Dividing this number (i.e., 11) by the total number of comparisons (i.e., 100)

EXHIBIT 5.5 ■ Data Matrix for Bootstrap MGA

Bootstrap Samples $p^{[2]}$	Bootstrap Samples $p^{[1]}$									
	0.357	0.226	0.318	0.281	0.372	0.318	0.296	0.308	0.415	0.272
0.494										
0.423										
0.324	X				X				X	
0.591										
0.698										
0.291	X		X		X	X	X	X	X	
0.509										
0.400									X	
0.526										
0.538										

Note: X indicates situations in which $p^{[1]} \geq p^{[2]}$.

yields the p value, which is 0.110 in our case. Therefore, we cannot conclude that the path coefficient p is significantly larger in group 1 than in group 2.

The bootstrap MGA involves a great number of comparisons of bootstrap estimates. For an initial analysis of group differences, we suggest drawing on 500 bootstrap samples, using the algorithm settings (e.g., no sign change option) as recommended by Hair, Hult, Ringle, & Sarstedt (2022). For the final analysis, use 1,000 bootstrap samples, which results in 1,000,000 comparisons for each parameter.

Finally, it is important to emphasize that the bootstrap MGA approach allows for testing only one-sided hypotheses (i.e., $p^{(1)} \geq p^{(2)}$). For instance, if the p value is smaller than a pre-defined probability of error of 0.05, we assume a significant result. However, a result can be significant for the other direction ($p^{(2)} \geq p^{(1)}$) as, for example, indicated by p values of 0.95 and above. When considering both directions in a two-sided test with a pre-defined probability of error of 0.05 (two-tailed), we consider p values of 0.025 and smaller as well as 0.975 as significant results. Note that for the ease of use, SmartPLS automatically creates a p value for the two-tailed test (i.e., we only need to check if the p values are smaller than 0.05 instead of considering p values of 0.025 and smaller, as well as 0.975 as significant results). Using the bootstrap MGA approach to test one- or two-sided hypotheses is not possible without limitations, as the bootstrap-based distribution is not necessarily symmetric. This characteristic clearly limits its applicability as researchers routinely draw on two-tailed tests—despite frequent concerns (e.g., Cho & Abe, 2012).

Permutation Test

The **permutation MGA** was originally developed by Chin (2003) and further substantiated by Chin and Dibbern (2010). As its name implies and analogous to its role in Step 2 of the MICOM procedure, the permutation MGA randomly exchanges observations between the data groups and re-estimates the model for each permutation. Computing the differences between the group-specific path coefficients per permutation enables testing whether these also differ in the population.

The permutation test entails the following six-step process:

1. Carry out group-specific estimations of the PLS path models to obtain path coefficient estimates for group 1 and group 2.

2. Compute the difference between the group-specific path coefficient estimates: $d = p^{(1)} - p^{(2)}$.

3. Produce random permutations of the data, meaning that the observations are randomly assigned to the groups. In formal terms, $n^{(1)}$ observations are drawn without replacement from the aggregate dataset and assigned to group 1. The remaining $n^{(2)}$ observations are assigned to group 2. Researchers should use a minimum number of 1,000 permutations.

4. Conduct group-specific estimations of the PLS path models for each permutation run u. For example, with 1,000 permutations, we obtain 1,000 model estimates each for group 1 (i.e., $p_u^{(1)}$) and for group 2 (i.e., $p_u^{(2)}$).

5. Compute the differences in the permutation run-specific path coefficient estimates: $d_u = p_u^{(1)} - p_u^{(2)}$.

6. Create the two-tailed 95% permutation-based confidence interval (assuming a 5% significance level). More specifically, sort the resulting d values in ascending order and determine the two values that separate (1) the 2.5% lowest from the 97.5% highest values (lower boundary) and (2) the 97.5% lowest from the 2.5% highest values (upper boundary). If the original difference d of the group-specific path coefficient estimates does not fall into the confidence interval, it is statistically significant. In addition, based on the permutation results, SmartPLS provides the p value of the difference d. A p value lower than 0.05 suggests that the difference d between the group-specific path coefficients does not fall into the 95% confidence interval and, thus, is statistically significant.

Comparing More Than Two Groups

Standard approaches to multigroup comparison in PLS-SEM have in common that they test the difference in the parameters between two groups. However, researchers frequently encounter situations in which they would like to compare a parameter (e.g., a path coefficient) across more than two groups. For example, comparing a specific path coefficient p across five groups requires conducting 10 comparisons (Exhibit 5.6). Not only do such complex analyses become increasingly time-consuming, but a more severe problem is associated with this approach, which is called **alpha inflation** (also referred to as **multiple testing problem**). This refers to the fact

EXHIBIT 5.6 ■ Pairwise Comparison of a Specific Path Coefficient Across Multiple Groups				
	Group 1	*Group 2*	*Group 3*	*Group 4*	*Group 5*
Group 1					
Group 2	$p^{(1)} - p^{(2)}$				
Group 3	$p^{(1)} - p^{(3)}$	$p^{(2)} - p^{(3)}$			
Group 4	$p^{(1)} - p^{(4)}$	$p^{(2)} - p^{(4)}$	$p^{(3)} - p^{(4)}$		
Group 5	$p^{(1)} - p^{(5)}$	$p^{(2)} - p^{(5)}$	$p^{(3)} - p^{(5)}$	$p^{(4)} - p^{(5)}$	

Note: p represents the specific path coefficient in the structural model; the comparison is across five groups; the superscript numbers in brackets indicate the group for which the specific coefficient was obtained.

that the more tests we conduct at a certain significance level, the more likely we are to claim a significant result when this is actually not the case (i.e., a Type I error). For example, assuming a significance level of 5% and making all possible pairwise comparisons as specified above, the overall probability of a Type I error (also referred to as the **familywise error rate**) is not 5% but 22.62% (Sarstedt & Mooi, 2019, Chapter 6). In other words, for more group comparisons, the familywise error rate quickly increases beyond the acceptable Type I error level (i.e., the acceptance of false positives) of, for example, 5%.

A standard approach for controlling for the familywise error rate is the **Bonferroni correction** (Holm, 1979). Instead of using a specific significance level in the test decision, the Bonferroni correction tests each individual hypothesis at a significance level of α/m, where α is the alpha level set by the researcher and m is the number of comparisons. For example, in case of five groups, there would be $m = 10$ comparisons, yielding a significance level of $0.05/10 = 0.005$ instead of 0.05. Further approaches exist—see Proschan and Brittain (2020) for an overview—but Klesel, Schubert, Niehaves, and Henseler's (2022) study has shown that these correction factors do not differ in their efficacy for controlling the familywise error rate. We therefore recommend applying the Bonferroni correction, which is easy to implement. Their simulation study has also disclosed the adverse effects of using too few bootstrapping samples or permutation runs when adjusting the results using, for example, the Bonferroni correction. While 1,000 bootstrap samples are sufficient in this regard, researchers should request 5,000 permutation runs when comparing more than two groups and controlling for alpha inflation.

Recommendations

The three approaches for comparing a single path coefficient have different statistical properties. From a conceptual perspective, the parametric MGA approach is problematic since it relies on distributional assumptions, which are inconsistent with PLS-SEM's nonparametric nature. However, this property does not translate into substantial differences in terms of false positives (i.e., higher Type I errors) when compared to nonparametric alternatives as shown in Klesel, Schubert, Niehaves, and Henseler's (2022) simulation study. Specifically, all three approaches exhibit very similar Type I error rates and levels of statistical power across the various simulation conditions considered in their study. Since the permutation MGA is slightly more powerful in detecting group differences compared to the parametric MGA (for both, equal and unequal variances) as well as the bootstrap MGA, and also maintains the Type I error rate when comparing two groups, we recommend its use for multigroup analyses in PLS-SEM. Exhibit 5.7 summarizes the rules of thumb for a combined multigroup analysis and invariance assessment in PLS-SEM.

In practical applications, researchers can apply different MGA approaches (e.g., permutation MGA, bootstrap MGA, and parametric MGA) to assess whether their

EXHIBIT 5.7 ■ Rules of Thumb for Invariance Assessment and Multigroup Analyses

Invariance assessment:

● Running an MGA requires partial measurement invariance, which is given when configural invariance and compositional invariance hold. If partial measurement invariance is confirmed and the composites have equal mean values and variances across the groups, there is full measurement invariance, which supports the pooled data analysis.

● For configural invariance assessment, check whether the measurement models employ the same indicators across the groups in terms of numbers, item content, and coding. Any type of data treatment (e.g., outliers or missing values) must be applied identically across the groups. Group-specific model estimations must draw on the same algorithm settings.

● Use 1,000 permutations (or more) for the assessment of compositional invariance and the equality of composite mean values and variances in Steps 2 and 3 of the MICOM procedure.

● Compositional invariance is established when the original correlations between the composite scores of groups 1 and 2 are not significantly different at a significance level of 5%.

● In exploratory research and when the sample size is small, reverting to the 1% significance level is more reasonable.

● Testing for the equality of the composites' mean values and variances across the groups requires assessing whether the original group differences fall inside the corresponding confidence intervals. If they do, the means and variances can be assumed to be equal.

Multigroup analyses:

● Preferably use the permutation approach when conducting an MGA using 1,000 permutations (or more).

● In case of pronounced differences in relative group sizes, consider drawing a random sample from the larger group to have a sample size that matches the one of the smaller group—provided the sample size allows for such a step.

● When comparing multiple groups, control for the familywise error rate using the Bonferroni correction. In this case, run the permutation test with 5,000 permutations (or more).

● Apply several MGA approaches to assess whether the results converge in a multimethod MGA approach.

results converge. If these alternative computations show that a particular parameter is significantly different between groups, this substantiates the results obtained via a **multimethod MGA approach**. On the contrary, considerable variation in results when drawing on alternative MGA approaches suggests that any conclusion regarding a significant difference should be considered with caution.

As a caveat, Klesel, Schubert, Niehaves, and Henseler's (2022) simulation study also shows that all approaches require sample sizes of 600 and higher to reliably detect even pronounced path coefficient differences of 0.2. Smaller differences of 0.1 require even sample sizes of more than 3,000 observations. Researchers should be cognizant of these requirements when conducting research in which they expect significant differences between the groups. Disclosing small differences in path coefficients requires relatively large sample sizes, which must be considered in the data collection stage. In addition, differences in the relative group sizes further diminish the approaches' power, albeit to a limited degree, unless the differences are pronounced (one-third in one group, two-thirds in the other group) and the data are nonnormally distributed. In such a setting, researchers need to randomly draw another sample from the large group that is similar in size to the smaller group and compare the two samples using the permutation MGA—provided the overall sample size allows for such an assessment.

CASE STUDY ILLUSTRATION

To illustrate the use of an MGA in conjunction with the assessment of measurement model invariance by means of the MICOM procedure, we draw on the **Corporate reputation model** in the **Example – Corporate reputation (advanced)** project. Rather than analyzing the aggregate dataset with 344 observations, we are interested in analyzing whether the effects in the model differ significantly for customers with prepaid cell phone plans from those with contract plans. The research study obtained data on customers with a prepaid plan (*servicetype* = 1; $n^{(1)}$ = 125) versus those with a contract plan (*servicetype* = 2; $n^{(2)}$ = 219), and we will compare these two groups.

When engaging in an MGA, it is important to ensure that the number of observations in each group meets the rules of thumb for minimum sample size requirements. As the maximum number of arrows pointing at a construct is eight, we would need at least a number of 80 observations per group, according to the 10 times rule. Following the more rigorous inverse square root method (Kock & Hadaya, 2018) and assuming a power level of 80%, a significance level of 5%, and a minimum path coefficient of 0.20, we would need 155 observations per group (Hair, Hult, Ringle, & Sarstedt. 2022, Chapters 1 & 5). At the same time, relying on the results of a power analysis (Hair, Hult, Ringle, & Sarstedt, 2022, Chapter 5), 54 observations per group are needed to detect R^2 values of around 0.25 at a significance level of 5% and a power level of 80%. Jointly considering these results, the group-specific sample sizes can be regarded sufficiently large.

In the first step, we have to define the grouping variable that SmartPLS uses to split the dataset. To do so, double-click on the dataset **Complete data [344]** in

EXHIBIT 5.8 ■   Generate Data Groups in SmartPLS

📋 Generate data groups   ✕

Cartesian product indicators

☐ $0 - serviceprovide.
☑ $1 - servicetype
☐ $2 - comp_1
☐ $3 - comp_2
☐ $4 - comp_3
☐ $5 - like_1
☐ $6 - like_2
☐ $7 - like_3
☐ $8 - cusl_1
☐ $9 - cusl_2

Name pattern

Group_$1

Minimum cases per group

54

Close   Apply

the **Example – Corporate reputation (advanced)** project. SmartPLS will open a new tab, which provides information on the dataset and its format (data view). When opening the data view, three options appear in the menu above the modeling window: **Add group, Generate groups**, and **Clear groups.** To define a new grouping variable, click on **Generate groups** and a new dialog box, which shows the available indicators in the dataset (Exhibit 5.8), will open. In the list, tick the box next to $1 – *servicetype*. The text field below labeled **Name pattern** shows the group name (**Group_$1**) with the default prefix *Group_* before the number of the grouping variable in the dataset. Retaining this default setting, the groups will initially be labeled *Group_1* and *Group_2*. Finally, we change the **Minimum cases per group** from the default value of 10 to 54 in our corporate reputation model example. As a result, SmartPLS generates groups based on the defined grouping variable only if the respective group contains at least 54 observations (otherwise the group is not formed). Close the dialog box by left-clicking on **Apply**.

The data group view shows the generated groups including the number of cases per group (Exhibit 5.9). The **Edit** option allows changing the group names. We change the label of Group_1.000 into *Prepaid plan* and the label of Group_2.000 into *Contract plan* and click on **Apply**. Use the **Navigation** menu on the left-hand side to return to the overview of **Indicators** with the descriptive statistics for the overall dataset.

We close the data view by clicking on **Back** in the menu bar. In the Workspace window on the left-hand side, double-click on **Corporate reputation model** in the **Example – Corporate reputation (advanced)** project. The corporate reputation model appears in the modeling window and the available indicators on the left-hand side of the screen. Above the indicator list appears the name of the model and the dataset selected for model estimation. Make sure to select **Complete data** [344].

To run the MICOM procedure, navigate to **Calculate → Permutation multigroup analysis** in the SmartPLS tool bar. A dialog box with alternative analysis options for the permutation method opens (Exhibit 5.10). In this dialog box, we first need to specify the groups to be compared. To do so, click on the menu next to **Group A** and select *Prepaid plan*. Next, click on the menu next to **Group B** and select *Contract plan*. Specify at least a number of **1,000 Permutations** and **Two tailed** under **Test type**. Since service type is known to affect the way customers relate to experiences, we assume a **Significance level** of **0.05**. Make sure to select fixed seed under **Random number generator** so that we obtain the same results when rerunning the random permutation method. Next, confirm that the box next to **Open report** has been ticked and click on **Start calculation**. SmartPLS now

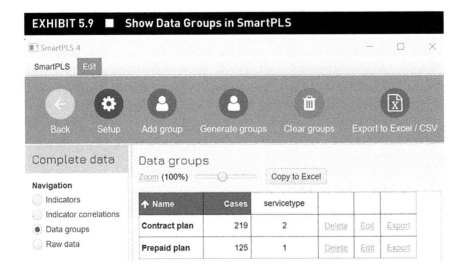

EXHIBIT 5.9 ■ Show Data Groups in SmartPLS

EXHIBIT 5.10   ■   Permutation Dialog Box

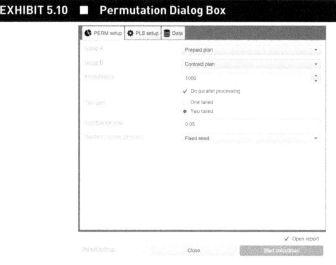

estimates the model for each group separately, using the standard PLS-SEM algorithm settings.

The PLS path models, the data treatment applied to both groups, and the algorithm settings used are identical for the group-specific model estimations, which is a necessary requirement for the establishment of configural invariance in Step 1 of the MICOM procedure. The following analyses address Step 2 and Step 3 of the MICOM procedure. After the permutation algorithm is completed, SmartPLS opens the **Results report** and displays the model with the results. In the option on top of the modeling window, we can select the results to be displayed in the modeling window. In the **Results report**, navigate to **Quality Criteria → MICOM**. The first results table labeled **MICOM – Step 2** (see Exhibit 5.11) shows the results of the compositional invariance assessment. Even though permutation is a random process, our results should look the same provided that we used the fixed seed option in the permutation settings. Selecting random seeds will produce different results, every time the permutation is run. However, the results should be very similar and not entail different conclusions. If the conclusions alternate from one permutation run to the next, this casts doubt on the establishment of compositional invariance.

The column **5.0%** shows the 5% quantile of the empirical distribution of c_u. Comparing the correlations c between the composite scores of group 1 and 2 (column **Original correlation**) with the 5% quantile reveals that the quantile is always smaller than (or equal to) the correlation c for all the constructs. This result is also supported by the **Permutation p values,** which in this case are always higher than 0.05 (and therefore colored in green), indicating that the correlations are not significantly lower than 1 (Exhibit 5.11). For example, the original correlation value of $ATTR$ is 0.993, which is higher than the 5% quantile (0.941); the corresponding p value of 0.691 is also

EXHIBIT 5.11 ■ **MICOM Step 2 in SmartPLS**

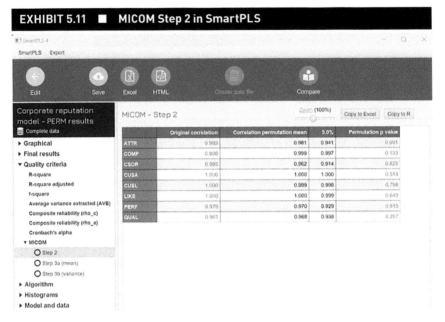

considerably larger than 0.05. Hence, the original correlation of *ATTR* is not significantly different from 1, which supports the conclusion that compositional invariance has been established for this construct. Similarly, we substantiate that compositional invariance has been established for all other multi-item constructs in the model. Note that we cannot run Step 2 for the single-item construct *CUSA* since its only indicator has a weight of 1 by design (i.e., in both groups). Thus, we can ignore *CUSA*'s *p* value of 0.514. To summarize, the results from Step 2 support partial measurement invariance. Thus, we can compare the standardized path coefficients across the groups by means of an MGA. To check whether full measurement invariance holds, click on the tabs **Step 3a (mean)** and **Step 3b (variance)** in the results report.

When clicking on **Step 3a (mean)**, the first two columns in Exhibit 5.12 show the mean differences between the composite scores as resulting from the original model estimation and the permutation procedure, respectively. The next two columns show the lower (**2.5%**) and upper (**97.5%**) boundaries of the 95% confidence interval of the scores' mean differences. As we can see, every confidence interval includes the original difference in mean values, indicating that there are no significant differences in the mean composite scores across the two groups. For example, for *ATTR*, the original difference in mean scores is –0.064, which is within the corresponding confidence interval with a lower boundary of –0.228 and an upper boundary of 0.203. The result in the column **Permutation p values** further supports this finding for *ATTR* and all other constructs in the PLS path model, as all the *p* values are considerably larger than 0.05 (and therefore colored in green).

Exhibit 5.13 shows the corresponding results for the composite scores' variances. Again, all the confidence intervals include the original value and all the *p* values are clearly larger than 0.05. We therefore conclude that all the composite mean values and

EXHIBIT 5.12 ■ MICOM Step 3a in SmartPLS

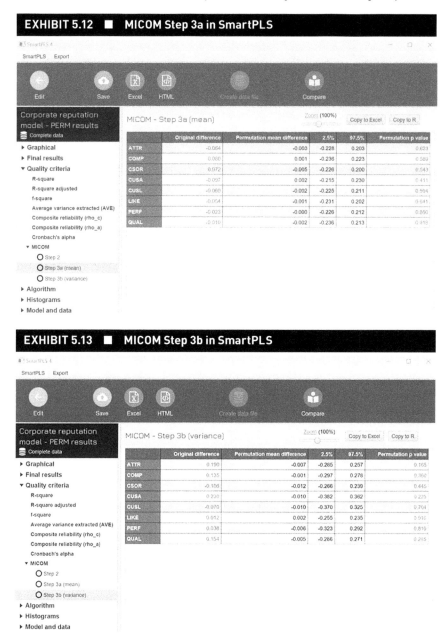

EXHIBIT 5.13 ■ MICOM Step 3b in SmartPLS

variances are equal, providing support for full measurement invariance. Exhibit 5.14 summarizes the results of this MICOM analysis.

In light of these results, we continue by examining the results of the MGA. Our assessment focuses on the permutation MGA, which we can access by going to **Final results → Path coefficients** in the SmartPLS permutation output. The first two columns in Exhibit 5.15 show the original path coefficients in the two groups. The following

EXHIBIT 5.14 ■ **Summary of the MICOM Results**

MICOM Step 1

Configural variance established? Yes

MICOM Step 2

Composite	Correlation c	5% quantile of the empirical distribution of c_u	p value	Compositional invariance established?
ATTR	0.993	0.941	0.691	Yes
COMP	0.998	0.997	0.133	Yes
CSOR	0.985	0.914	0.825	Yes
CUSA	1.000	1.000	0.514	Yes
CUSL	1.000	0.998	0.798	Yes
LIKE	1.000	0.999	0.649	Yes
PERF	0.979	0.929	0.615	Yes
QUAL	0.961	0.938	0.267	Yes

MICOM Step 3a and Step 3b

Composite	Difference of the composite's mean value	95% confidence interval	p value	Equal mean values?
ATTR	-0.064	[-0.228; 0.203]	0.623	Yes
COMP	0.060	[-0.236; 0.223]	0.589	Yes
CSOR	0.072	[-0.226; 0.200]	0.543	Yes
CUSA	-0.097	[-0.215; 0.230]	0.411	Yes
CUSL	-0.060	[-0.225; 0.211]	0.594	Yes
LIKE	-0.054	[-0.231; 0.202]	0.641	Yes
PERF	-0.023	[-0.226; 0.212]	0.850	Yes
QUAL	-0.010	[-0.236; 0.213]	0.918	Yes

EXHIBIT 5.14 ■ Summary of the MICOM Results *(Continued)*

Composite	Logarithm of the composite's variances ratio	95% confidence interval	*p* value	Equal variances?
ATTR	0.190	[−0.285; 0.257]	0.165	Yes
COMP	0.135	[−0.297; 0.278]	0.360	Yes
CSOR	−0.106	[−0.266; 0.239]	0.445	Yes
CUSA	0.230	[−0.382; 0.362]	0.225	Yes
CUSL	−0.070	[−0.370; 0.325]	0.704	Yes
LIKE	0.012	[−0.255; 0.235]	0.916	Yes
PERF	0.038	[−0.323; 0.292]	0.810	Yes
QUAL	0.154	[−0.286; 0.271]	0.265	Yes

EXHIBIT 5.15 ■ Permutation MGA Results in SmartPLS

Path coefficients

Zoom (100%)   Copy to Excel   Copy to R

	Original (Prepaid plan)	Original (Contract plan)	Original difference	Permutation mean difference	2.5%	97.5%	Permutation p value
ATTR -> COMP	0.163	0.049	0.114	-0.001	-0.226	0.218	0.337
ATTR -> LIKE	0.245	0.133	0.112	-0.002	-0.265	0.259	0.416
COMP -> CUSA	0.214	0.108	0.106	-0.004	-0.297	0.274	0.468
COMP -> CUSL	0.136	-0.062	0.197	0.005	-0.224	0.225	0.083
CSOR -> COMP	-0.026	0.091	-0.118	-0.003	-0.229	0.225	0.329
CSOR -> LIKE	0.096	0.199	-0.103	-0.003	-0.233	0.227	0.377
CUSA -> CUSL	0.599	0.440	0.159	-0.005	-0.181	0.161	0.069
LIKE -> CUSA	0.365	0.476	-0.111	0.000	-0.261	0.254	0.391
LIKE -> CUSL	0.205	0.424	-0.220	0.003	-0.217	0.224	0.052
PERF -> COMP	0.210	0.305	-0.095	0.002	-0.262	0.262	0.508
PERF -> LIKE	-0.002	0.172	-0.174	0.001	-0.278	0.269	0.218
QUAL -> COMP	0.535	0.418	0.117	0.007	-0.258	0.257	0.411
QUAL -> LIKE	0.501	0.336	0.165	0.008	-0.230	0.273	0.227

columns include the path coefficient differences in the original dataset (**Original difference**) and the average differences across all permutation runs (**Permutation mean difference**). The final columns show the lower (**2.5%**) and upper (**97.5%**) boundaries of the 95% confidence interval of the permutation-based differences and the corresponding *p* value (**Permutation p value**). As can be seen, neither of the structural model relationships is significantly different between the two groups. All the *p* values are larger than 0.05 as indicated by the red font color. Assuming a more liberal significance level of 10% produces significant differences in the path coefficients between *COMP* and *CUSL, CUSA* and *CUSL*, as well as *LIKE* and *CUSL*. More precisely, the effect between *COMP* and *CUSL* is significantly different (*p* value = 0.083) between customers with a prepaid plan ($p^{(1)}$ = 0.136) and those with a contract plan ($p^{(2)}$ = −0.062). Similarly, the relationship between *CUSA* and *CUSL* is significantly different (*p* value = 0.069) among customers with a prepaid plan ($p^{(1)}$ = 0.599) versus those with a contract plan ($p^{(2)}$ = 0.440). Finally, the effect of *LIKE* on *CUSL* is significantly different (p value = 0.052) among customers with a prepaid plan ($p^{(1)}$ = 0.205) versus those with a contract plan ($p^{(2)}$ = 0.424).

To further analyze group-specific effects, we run the bootstrap MGA approach to assess the results' stability. Click on **Back** and select **Calculate → Bootstrap multigroup analysis** from the tool bar. In the dialog box that opens (Exhibit 5.16, Panel A), we again need to specify the groups to be compared. To do so, tick the box next to *Prepaid plan* under **Groups A** and select *Contract plan* under **Groups B**. In the other tabs, we can specify settings related to the PLS-SEM algorithm, the bootstrapping procedure, missing value treatment, and variable weighting. Use the standard settings for all the options as described in, for example, Hair, Hult, Ringle, and Sarstedt (2022). Particular attention needs to be paid to the choice of the number of **subsamples** in the **BT setup** tab. As the bootstrap MGA involves piecewise comparisons of all group-specific boot-strap estimates, we recommend limiting the number of subsamples to **1,000**. This number yields 1,000 · 1,000 = 1,000,000 comparisons of group-specific bootstrap results. In addition, we should select the percentile bootstrap under **Confidence interval method** (Exhibit 5.16, Panel B) as this approach produces the lowest coverage error (Hair, Hult, Ringle, & Sarstedt, 2022, Chapter 5). Before clicking on **Start calculation**, make sure to tick the box next to **Open report**. In the results report that opens, navigate to **Final Results → Path Coefficients** and select the **Bootstrap MGA** tab to access the results of this MGA approach (Exhibit 5.17).

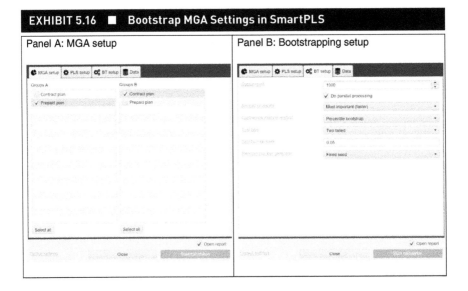

EXHIBIT 5.16 ■ Bootstrap MGA Settings in SmartPLS

Panel A: MGA setup

Panel B: Bootstrapping setup

EXHIBIT 5.17 ■ Bootstrap MGA Results in SmartPLS

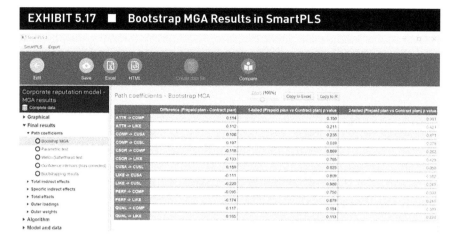

The bootstrap MGA represents a one-tailed test with the *p* values indicating whether the path coefficient is significantly larger in group 1 (prepaid plan) than in group 2 (contract plan). However, a result can be significant for the other direction in that the second group (contract plan) is significantly larger than the first group (prepaid plan). Hence, when considering a pre-defined probability of error of **0.05** (two-tailed), we assume *p* values of smaller than 0.025 as well as larger than 0.975 as indicative of significant differences between the group-specific path coefficients. The results column **1-tailed (Prepaid plan vs. Contract plan) p value** in Exhibit 5.17 shows that neither of the *p* values is smaller than 0.025. However, the *LIKE* → *CUSL* relationship has a *p* value of 0.980, which is larger than 0.975. Hence, we conclude that the difference of –0.220 between the two group-specific model estimations is significant. This finding is also supported by the results column **2-tailed (Prepaid plan vs. Contract plan) p value**, in which we only need to check if the *p* values are smaller than 0.05.

Finally, Exhibit 5.18 presents the results of the parametric test, which we can access by clicking **Final Results → Path Coefficients → Parametric test** in the results report. Here, we show the results of the parametric MGA (equal) since the MICOM procedure revealed that the variances do not differ across groups. However, the parametric MGA (unequal) results, which we can find in the tab labeled **Welch-Satterthwait test** of the results report, are very similar.

To summarize, we find that the various MGA methods do not yield different conclusions as shown in Exhibit 5.19. Even though we recommend using the permutation test, running a multimethod MGA approach by applying several approaches provides additional confidence in the conclusions drawn from the analysis.

EXHIBIT 5.18 ■ Parametric MGA Results

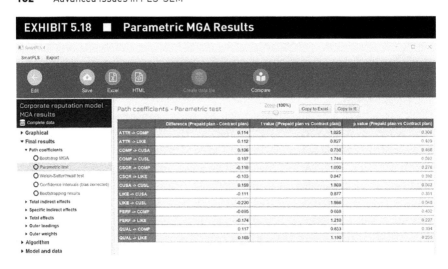

EXHIBIT 5.19 ■ PLS Multigroup Results Across Methods

Path Coefficient	Permutation Test	Bootstrap MGA	Parametric Test	Welch-Satterthwaite t Test
ATTR → COMP				
ATTR → LIKE				
COMP → CUSA				
COMP → CUSL	X	X	X	X
CSOR → COMP				
CSOR → LIKE				
CUSA → CUSL	X	X	X	X
LIKE → CUSA				
LIKE → CUSL	X	X	X	X
PERF → COMP				
PERF → LIKE				
QUAL → COMP				
QUAL → LIKE				

Note: X indicates significant difference (*p* < 0.10) of path coefficients across the groups.

SUMMARY

- **Realize the importance of considering heterogeneity in PLS-SEM.**
Applications of PLS-SEM are usually based on the assumption that the data represent a homogeneous population. In such cases, the estimation of a global PLS path model represents all observations. In many real-world applications, however, the assumption of sample homogeneity is unrealistic because respondents are likely to be heterogeneous in their perceptions and evaluations of latent phenomena. While the consideration of heterogeneity is promising from a practical and theoretical perspective to learn about differences between groups of respondents, it is oftentimes necessary, but also challenging, to obtain valid results. For example, when there are significant differences between path coefficients across groups, an analysis on the aggregate data level could cancel out group-specific effects. The results of such an analysis would likely be seriously misleading and render erroneous recommendations to managers and researchers. Researchers are therefore well advised to consider the issue of heterogeneous data structures in their modeling efforts.

- **Explain the difference between observed and unobserved heterogeneity.**
Heterogeneity can be observed in that differences between two or more groups of data relate to observable characteristics, such as gender, age, or country of origin. Researchers can use these observable characteristics to partition the data into separate groups of observations and carry out group-specific PLS-SEM analyses. Moreover, heterogeneity can be unobserved in that it does not depend on one specific observable characteristic or combinations of several characteristics. To identify and treat unobserved heterogeneity, researchers can draw on a range of latent class techniques.

- **Comprehend the concept of measurement model invariance and its assessment in PLS-SEM.** Group comparisons are valid if measurement invariance has been established. Thereby, researchers ensure that group differences in model estimates do not result from the distinctive content and/or meanings of the constructs across groups. When measurement invariance is not demonstrated, any conclusions about differences in model relationships are questionable. The measurement invariances of composites (MICOM) procedure represents a useful tool in PLS-SEM. The procedure comprises three steps that test different aspects of measurement invariance: (1) configural invariance (i.e., equal parameterization and way of estimation), (2) compositional invariance (i.e., equal indicator weights), and (3) equality of composite mean values and variances.

- **Understand PLS-SEM multigroup analysis and execute it in SmartPLS.**
  The MGA enables researchers to test whether differences between group-specific path coefficients are statistically significant. Researchers have proposed different approaches to the MGA, with the t test-based parametric MGA approach being most frequently applied. Alternatives such as the permutation MGA and the bootstrap MGA do not rely on distributional assumptions. In comparison, the permutation MGA has particularly advantageous statistical properties, and we recommend this method when analyzing the difference of parameters across two groups. SmartPLS enables the application of the MICOM procedure to measurement invariance assessment and the different MGA approaches via its permutation MGA and bootstrap MGA algorithms. However, each method requires decisions by the researcher about the setting of the parameters. Running a multimethod MGA approach by applying several approaches provides additional confidence in the conclusions drawn from the analysis.

REVIEW QUESTIONS

1. What is the difference between observed and unobserved heterogeneity?

2. Explain the three steps of the MICOM procedure.

3. Why would you run an MGA in PLS-SEM?

4. Which approaches to MGA are available in PLS-SEM? Discuss their advantages and disadvantages.

CRITICAL THINKING QUESTIONS

1. Why is the consideration of heterogeneity so important when analyzing PLS path models?

2. Critically comment on the following statement: Measurement invariance is not an issue in PLS-SEM because of the method's dual focus on explanation and prediction.

3. What are the implications for your analysis when full measurement invariance holds?

4. Explain the conceptual differences between an MGA and an analysis of interaction effects.

KEY TERMS

Alpha inflation

Bonferroni correction

Bootstrap MGA

Bootstrap samples

Bootstrapping

Cluster analysis

Common factor models

Compositional invariance

Confidence interval

Configural invariance

Equality of composite mean values and variances

Familywise error rate

Finite mixture partial least squares (FIMIX-PLS)

Full measurement invariance

k-means clustering

Latent class techniques

Measurement equivalence

Measurement invariance

Measurement invariance of composite models (MICOM) procedure

Multigroup analysis (MGA)

Multimethod MGA approach

Multiple testing problem

Nonparametric distance-based test

Observed heterogeneity

Parametric MGA

Parametric MGA (equal)

Parametric MGA (unequal)

Partial measurement invariance

Permutation

Permutation MGA

Prediction-oriented segmentation in PLS-SEM (PLS-POS)

Unobserved heterogeneity

Welch-Satterthwaite t test

SUGGESTED READINGS

Becker, J.-M., Rai, A., Ringle, C. M., & Völckner, F. (2013). Discovering unobserved heterogeneity in structural equation models to avert validity threats. *MIS Quarterly, 37*(3), 665–694.

Cheah, J.-H., Amaro, S., & Roldán, J. L. (2023). Multigroup analysis of more than two groups in PLS-SEM: A review, illustration, and recommendations. *Journal of Business Research, 156*, 113539.

Chin, W. W., & Dibbern, J. (2010). A permutation based procedure for multi-group PLS analysis: Results of tests of differences on simulated data and a cross cultural analysis of the sourcing of information system services between Germany and the USA. In V. Esposito Vinzi, W. W. Chin, J. Henseler, & H. Wang (Eds.), *Handbook of partial least squares: Concepts, methods and applications in marketing and related fields* (pp. 171–193). Springer.

Henseler, J., Ringle, C. M., & Sarstedt, M. (2016).Testing measurement invariance of composites using partial least squares. *International Marketing Review, 33*(3), 405–431.

Klesel, M., Schuberth, F., Henseler, J., & Niehaves, B. (2019). A test for multigroup comparison in partial least squares path modeling. *Internet Research, 29*(3), 464–477.

Klesel, M., Schuberth, F., Niehaves, B., & Henseler, J. (2022). Multigroup analysis in information systems research using PLS-PM: A systematic investigation of approaches. *The Data Base for Advances in Information Systems, 53*(3), 26–48.

Matthews, L. (2017). Applying multigroup analysis in PLS-SEM: A step-by-step process. In R. Noonan & H. Latan (Eds.), *Partial least squares structural equation modeling: Basic concepts, methodological issues and applications* (pp. 219–243). Springer.

Sarstedt, M., Henseler, J., & Ringle, C. M. (2011). Multi-group analysis in partial least squares (PLS) path modeling: Alternative methods and empirical results. *Advances in International Marketing, 22*, 195–218.

Schlägel, C., & Sarstedt, M. (2016). Assessing the measurement invariance of the four-dimensional cultural intelligence scale across countries: A composite model approach. *European Management Journal, 34*(6), 633–649.

6 MODELING UNOBSERVED HETEROGENEITY

LEARNING OUTCOMES

1. Comprehend the FIMIX-PLS and PLS-POS approaches for identifying and treating unobserved heterogeneity.

2. Understand how to use FIMIX-PLS and PLS-POS jointly in a latent class analysis.

3. Appreciate how to execute a FIMIX-PLS analysis using SmartPLS.

4. Know how to use SmartPLS to run a PLS-POS analysis.

CHAPTER PREVIEW

As described in Chapter 5, researchers routinely use observable characteristics to partition data into groups and estimate separate models, thereby accounting for **observed heterogeneity**. However, the sources of heterogeneity in the data can hardly be fully known a priori. Consequently, situations arise in which differences related to **unobserved heterogeneity** prevent finding accurate results since the analysis at the aggregate data level masks group-specific effects. Failure to consider heterogeneity can limit the validity of PLS-SEM results. Hence, identifying and—if necessary—treating unobserved heterogeneity is of crucial importance when using PLS-SEM. On the contrary, if unobserved heterogeneity does not substantially affect the results, researchers can analyze the data on the aggregate level, which is advantageous for the generalization of their conclusions.

In this chapter, we learn about two methods, finite mixture PLS (FIMIX-PLS) and prediction-oriented segmentation in PLS (PLS-POS) that enable researchers to identify and treat unobserved heterogeneity in PLS path models. Based on a description of the methods' basic working principles, we explain how they can be jointly used to identify the number of latent segments to retain from the data and estimate segment-specific path models. We also explain how to reproduce the latent segments by means of observable and managerially meaningful variables to turn the initial results into actionable understanding. The chapter concludes with a practical application of the methods drawing on our corporate reputation example and the SmartPLS software.

167

Uncovering Unobserved Heterogeneity in PLS Path Models

The usual way of treating heterogeneous data structures draws on standard **cluster analysis** methods such as **k-means clustering** (Sarstedt & Mooi, 2019, Chapter 9). In the context of PLS-SEM, we could use k-means clustering on the indicator data or constructs scores obtained from an initial analysis of the overall dataset to form groups of data. The partition this analysis produces can then be used as input for group-specific PLS-SEM estimations. While easy to apply, a major drawback of such an approach is that these methods do not account for the measurement and structural model relationships specified by the researcher. However, it is exactly these relationships that are likely responsible for some of the group differences. At the same time, prior research has shown that traditional clustering approaches perform poorly in identifying group differences in model relationships such as group-specific path coefficients (Sarstedt & Ringle, 2010).

In light of these limitations, methodological research in PLS-SEM has proposed a multitude of specific methods to identify and treat unobserved heterogeneity, commonly referred to as **latent class techniques** or **response-based segmentation techniques**. These techniques have proven very useful to identify unobserved heterogeneity and partition the data accordingly. Alternatively, latent class techniques can ascertain that unobserved heterogeneity does not influence the results, supporting an analysis of a single model based on the aggregate data level. Research has brought forward a multitude of different techniques that draw on, for example, genetic algorithms (Ringle, Sarstedt, & Schlittgen, 2014; Ringle, Sarstedt, Schlittgen, & Taylor, 2013), weighted least squares (Schlittgen, Ringle, Sarstedt, & Becker, 2016), reduced k-means clustering (Fordellone & Vichi, 2020), and other approaches to identify unobserved heterogeneity in a PLS-SEM framework—see Fordellone and Vichi (2020) as well as Hair, Sarstedt, Matthews, and Ringle (2016) for reviews of the various latent class approaches.

The first and most prominent latent class technique is **finite mixture partial least squares** (**FIMIX-PLS**; Hahn, Johnson, Herrmann, & Huber, 2002; Sarstedt, Becker, Ringle, & Schwaiger, 2011). As indicated by its name, the approach relies on the finite mixture models concept, which assumes that the overall population is a mixture of group-specific density functions. The aim of FIMIX-PLS is to disentangle the overall mixture distribution and estimate parameters (e.g., the path coefficients) of each group in a regression framework (i.e., mixture regressions; Becker, Ringle, Sarstedt, & Völckner, 2015; Wedel & Kamakura, 2000). Exhibit 6.1 shows an example of a mixture distribution that FIMIX-PLS aims to separate into segment-specific distributions.

FIMIX-PLS follows two steps. In the first step, the standard PLS-SEM algorithm is run on the full set of data to obtain the scores of all the constructs in the model. These construct scores then serve as input for a series of mixture regression analyses in the second step. The mixture regressions allow for the simultaneous probabilistic classification of observations into groups and the estimation of group-specific path

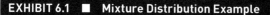

EXHIBIT 6.1 ■ Mixture Distribution Example

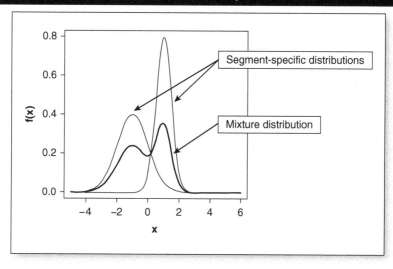

Source: Adapted from Hair, Sarstedt, Matthews, & Ringle (2016).

coefficients. While this procedure requires that the researcher explicitly defines the number of segments, FIMIX-PLS offers a range of statistical measures that provide an indication of how many segments likely underlie the data. This is a clear advantage of FIMIX-PLS as other PLS-SEM-based latent class techniques offer practically no insight in this respect (Sarstedt, Becker, Ringle, & Schwaiger, 2011).

Simulation studies show that FIMIX-PLS reliably reveals the existence of heterogeneity in PLS path models and correctly indicates the appropriate number of segments to retain from the data (Sarstedt, Becker, Ringle, & Schwaiger, 2011). At the same time, however, FIMIX-PLS is clearly limited in terms of correctly identifying the underlying segment structure, as defined by the group-specific path coefficients (Ringle, Sarstedt, Schlittgen, & Taylor, 2013; Ringle, Sarstedt, & Schlittgen, 2014), especially when the path model includes formative measures (Becker, Rai, Ringle, & Völckner, 2013). Furthermore, FIMIX-PLS is only capable of capturing heterogeneity in the structural model relationships and cannot account for heterogeneity in the measurement models, which limits its usefulness for empirical research settings.

To overcome FIMIX-PLS's limitations, a range of other alternatives have been proposed. A very promising approach is Becker, Rai, Ringle, and Völckner's (2013) **prediction-oriented segmentation in PLS-SEM (PLS-POS)**, which follows a fundamentally different approach than FIMIX-PLS. Rather than defining heterogeneity at a distributional level (Exhibit 6.1), PLS-POS gradually reallocates observations from one segment to the other with the application of a goal criterion, which is the maximization of the explained variance provided by the segmentation solution. More specifically, PLS-POS computes each observation's distance to its own segment as well as other segments to decide on its group membership. When an observation has the shortest distance to its own segment, it remains in the current segment. Otherwise, the method (re-)assigns the observations to the alternative segment that is the shortest distance away. The algorithm repeatedly conducts this procedure for all the observations in the dataset, thereby systematically improving the segment solution toward the goal criterion (i.e., maximization of the amount of explained variance across the segments). Becker, Rai, Ringle, and Völckner's (2013) simulation study shows that PLS-POS reliably identifies segment structures and outperforms alternative segmentation techniques. However, PLS-POS does not offer concrete recommendations regarding the number of segments to consider in the analysis.

Against this background, and because of these approaches' joint implementation in SmartPLS 4 (Ringle, Wende, & Becker, 2022), researchers benefit from using a combination of FIMIX-PLS and PLS-POS. The first step involves running FIMIX-PLS to identify the appropriate number of segments to retain from the data. The FIMIX-PLS solution then serves as a starting point for running PLS-POS to improve the FIMIX-PLS solution. PLS-POS also identifies heterogeneity in the measurement models. However, since the identified segments are—by definition—latent, the model estimates offer only limited guidance regarding the relationships to be expected in the real world. The final step, therefore, involves reproducing the latent segments by means of observable and managerially meaningful variables in order to convert the initial results into actionable understanding. Exhibit 6.2 summarizes the steps in the

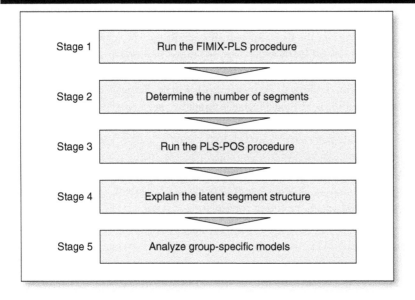

EXHIBIT 6.2 ■ A Systematic Procedure for Jointly Applying FIMIX-PLS and PLS-POS

Stage 1 — Run the FIMIX-PLS procedure

Stage 2 — Determine the number of segments

Stage 3 — Run the PLS-POS procedure

Stage 4 — Explain the latent segment structure

Stage 5 — Analyze group-specific models

systematic joint application of FIMIX-PLS and PLS-POS. Building on Hair, Sarstedt, Matthews, and Ringle (2016) and Sarstedt, Ringle, and Gudergan (2016), we discuss each step in greater detail in the following sections.

Step 1: Run the FIMIX-PLS Procedure

Running the FIMIX-PLS procedure requires the researcher to make several choices in the algorithm settings. The model estimation process in FIMIX-PLS follows the likelihood principle, which asserts that all of the evidence in a sample that is relevant for the model parameters is contained in the likelihood function. This likelihood function is maximized by using the **expectation-maximization (EM) algorithm**. The EM algorithm alternates between performing an expectation (E) step and a maximization (M) step. The E step creates a function for the expectation of the log-likelihood, which is evaluated using the current estimate of the parameters. The M step computes parameters by maximizing the expected log-likelihood identified in the E step. The E and M steps are successively applied until the results stabilize. Stabilization is reached when there is no substantial improvement in the (log) likelihood value from one iteration to the next. A threshold value of 10^{-7} is recommended as a stop criterion to ensure that the algorithm converges at reasonably low levels of iterative changes in the log-likelihood values. When the stop criterion is set very low, the FIMIX-PLS algorithm may not converge within a reasonable time. Therefore, the researcher also needs to specify a maximum number of iterations after which the algorithm will automatically terminate. Specifying a maximum number of 5,000 iterations usually ensures a sound balance between an acceptable computational running time and obtaining sufficiently precise results.

Using the EM algorithm for model estimation is attractive because it is efficient and always converges to a predefined number of segments. However, convergence might occur based on a local optimum solution, which differs from the global optimum solution (Steinley, 2003). To investigate the possible occurrence of a local optimum solution, researchers should run FIMIX-PLS multiple times. The initialization in FIMIX-PLS occurs randomly, which means each time FIMIX-PLS is initiated, the algorithm uses different randomly selected starting values for parameter estimation. Generally, however, the results of multiple FIMIX-PLS computations will be very similar. However, if the results are not similar, then a local optimum has occurred and the solution that is not consistent with the other solutions should be discarded. In line with simulation study results of the technique (Sarstedt, Becker, Ringle, & Schwaiger, 2011), we suggest using at least 10 repetitions of the FIMIX-PLS approach and choosing the solution with the best log-likelihood value—more repetitions (e.g., 25) are certainly more beneficial in securing a (close to the) global optimum solution. Another consequence of the algorithm's random nature is that the numbering of the segments is not determinate. That is, the results for one segment might appear in a different segment when FIMIX-PLS is run again. This characteristic is commonly referred to as **label switching** (McLachlan & Peel, 2000).

A further important consideration when running FIMIX-PLS involves the treatment of missing values. Kessel, Ringle, and Sarstedt (2010) have shown that just 5% missing values in one variable can cause severe problems in a FIMIX-PLS analysis when replacing them with the overall sample mean of that indicator's valid values (i.e., mean value replacement). In this case, the missing values treatment option creates a set of common scores, which FIMIX-PLS identifies as a distinct homogeneous segment. As a consequence, the number of segments will likely be overspecified and observations that truly belong to other segments will be forced into this artificially generated one. Therefore, mean value replacement must not be used in a FIMIX-PLS context, even if there are only few missing values in the dataset. Instead, researchers should remove all cases that include missing values in any of the indicators used in the model from the analysis (i.e., casewise deletion). While this approach also has its problems, particularly when values are missing at random (Sarstedt & Mooi, 2019, Chapter 5.4.5), it avoids the generation of an artificial segment as is the case with mean value replacement and other imputation methods, such as EM imputation and regression imputation.

Finally, the FIMIX-PLS algorithm needs to be run for alternating numbers of segments, starting with the one-segment solution. As the number of segments is a priori unknown, researchers must compare the solutions with the different segment numbers in terms of their statistical adequacy and interpretability. The range of possible segment numbers depends on the interplay between the sample size and the minimum sample size requirements to reliably estimate the given model. For example, when analyzing a dataset with 200 observations and the requirement of a minimum segment sample size of 50, it is not reasonable to run FIMIX-PLS with more than four segments. It is therefore imperative to consider model-specific minimum segment sample size requirements, as documented, for example, in Hair, Hult, Ringle, and Sarstedt (2022, Chapter 1), before defining a range of segment solutions to consider in the FIMIX-PLS analysis. The theoretical maximum number of segments to consider is determined by the largest integer, which is obtained by dividing the sample size n by the minimum sample size. However, since it is highly unlikely that the observations are evenly distributed across the segments, especially when the upper bound is high, considering a lower number of segments is generally preferred (Sarstedt, Ringle, & Hair, 2017). This is also why researchers need to check the sizes of the segments produced by FIMIX-PLS in Step 2 of the analysis.

Step 2: Determine the Number of Segments

A fundamental challenge with the application of FIMIX-PLS is determining the number of groups to retain from the data. Identifying a suitable number of groups is crucial, since managerial decisions are based on this result. As Becker, Ringle, Sarstedt, and Völckner (2015, p. 644) note, "a misspecified number of segments results in under- or oversegmentation, which can lead to inaccurate management decisions regarding, for example, customer targeting, product positioning, or determining the optimal marketing mix." FIMIX-PLS enables researchers to compute likelihood-based **information criteria** (also referred to as **model selection criteria**), which provide an indication of how many segments to retain from the data. Information criteria simultaneously consider the fit (i.e., the likelihood) of a model and the number of parameters used to achieve that fit. The information criteria denote, therefore, a penalized likelihood function (i.e., the negative likelihood plus a penalty term, which increases with the number of segments; Sarstedt, 2008). The smaller the value of a certain information criterion, the better the segmentation solution. Prominent examples of information criteria include **Akaike's information criterion** (**AIC**; Akaike, 1973), **modified AIC with factor 3** (**AIC₃**; Bozdogan, 1994), **consistent AIC** (**CAIC**; Bozdogan, 1987), and **Bayesian information criterion** (**BIC**; Schwarz, 1978). For a formal presentation of these criteria see, for example, Sarstedt, Becker, Ringle, and Schwaiger (2011).

Information criteria are not scaled within a certain range of values (e.g., between 0 and 1). Rather, the criteria might take values in the 100s or 1,000s, depending on the starting point of the FIMIX-PLS algorithm, which is set randomly. Importantly, however, each criterion's values can be compared across different solutions with varying segment numbers. Therefore, the researcher needs to examine several solutions with alternating numbers of segments and select the model that minimizes a particular information criterion.

Sarstedt, Becker, Ringle, and Schwaiger (2011) have evaluated the efficacy of different information criteria in FIMIX-PLS across a broad range of data and model constellations. Their results demonstrate that researchers should jointly consider AIC₃ and CAIC, or, alternatively, AIC₃ and BIC. Whenever these sets of criteria indicate the same number of segments, this result likely suggests the appropriate number of segments. The **modified AIC with factor 4** (**AIC₄**; Bozdogan, 1994) and BIC generally perform well, while other criteria exhibit a pronounced overestimation tendency. This holds especially for AIC, which often overspecifies the correct number of segments by three or more segments. Still other criteria, such as **minimum description length 5** (**MDL₅**; Liang, Jaszak, & Coleman, 1992), show pronounced underestimation tendencies. Researchers can use this information to determine a certain range of reasonable segment numbers. For example, when AIC indicates a five-segment solution, retaining a smaller number of segments seems warranted. Exhibit 6.3 provides an overview of selected information criteria and highlights their performance in the context of FIMIX-PLS.

Information criteria are not a "silver bullet" to determine the most suitable number of segments in FIMIX-PLS, since information criteria do not provide an indication of how well separated the segments are. For this reason, researchers should consider the complementary use of entropy-based measures, such as the **normed entropy statistic** (**EN**; Ramaswamy, DeSarbo, Reibstein, & Robinson, 1993). The EN uses the observations' segment membership probabilities to indicate whether the partition is reliable. The more observations exhibit high segment membership probabilities, the more clear-cut their segment affiliation will be. The EN ranges between 0 and 1, with higher

EXHIBIT 6.3 ■	**Selected Information Criteria and Their Performance in FIMIX-PLS**	

Abbreviation	Criterion Name	Performance in FIMIX-PLS
AIC	Akaike's Information Criterion	● Weak performance ● Strong tendency to overestimate the number of segments ● Can be used to determine the upper limit of reasonable segmentation solutions
AIC$_3$	Modified AIC with Factor 3	● Fair to good performance ● Tends to overestimate the number of segments ● Works well in combination with CAIC and BIC
AIC$_4$	Modified AIC with Factor 4	● Good performance ● Similar tendencies to over- and underestimate the number of segments
BIC	Bayesian Information Criterion	● Good performance ● Tends to underestimate the number of segments ● Should be considered jointly with AIC$_3$
CAIC	Consistent Akaike's Information Criterion	● Good performance ● Tends to underestimate the number of segments ● Should be considered jointly with AIC$_3$
MDL$_5$	Minimum Description Length 5	● Weak performance ● Strong tendency to underestimate the number of segments ● Can be used to determine the lower limit of reasonable segmentation solutions.

Source: Adapted from Hair, Sarstedt, Matthews, & Ringle (2016, p. 70).

values indicating a better quality partition. Prior research provides evidence that EN values above 0.50 permit a clear-cut classification of data into the predetermined number of segments (e.g., Ringle, Sarstedt, & Mooi, 2010).

When deciding on the number of segments to retain, it is particularly important to keep in mind that the EM algorithm always converges to the prespecified number of segments. The result can be, however, that FIMIX-PLS forces a small subset of data into extraneous segments, simply because the researcher selected a number of segments, which is too high. Such extraneous segments account for only a marginal portion of heterogeneity in the overall dataset and are usually too small to ensure valid group-specific results (Rigdon, Ringle, & Sarstedt, 2010). Therefore, in addition to information criteria and the EN, the researcher should carefully consider the segment sizes produced by FIMIX-PLS. If the analysis yields an extraneous segment that is too small to warrant valid analysis, the researcher should consider reducing the number of segments or dropping this segment and focusing on the analysis and interpretation of the other, larger segments.

Finally, it is important to note that a purely data-driven approach provides only rough guidance regarding the number of segments to select. Heuristics such as information criteria and the EN have limitations because they are sensitive to data and model characteristics. For example, research results by Becker, Ringle, Sarstedt, & Völckner (2015) suggest that even low levels of collinearity in the structural model can have adverse consequences for the information criteria's performance. FIMIX-PLS is an exploratory tool and should be treated as such. Consequently, any decision regarding the number of segments should take practical considerations into account (e.g., Sarstedt, Schwaiger, & Ringle, 2009). For example, researchers might have a priori knowledge or a theory that can support their reasoning. Likewise, the number of segments must be small enough to ensure parsimony and manageability, but each segment should also be large enough to warrant strategic attention (Sarstedt & Mooi, 2019, Chapter 9).

Exhibit 6.4 shows a decision tree for determining the number of segments in FIMIX-PLS. The decision framework identifies two situations that call for using the aggregate data level results (i.e., no segmentation). First, if AIC_3 and CAIC indicate the same number of segments, but the EN statistic or the relative segment sizes do not offer support for this solution, researchers should revert to the aggregate results. Alternatively, if the AIC (upper boundary) and MDL_5 (lower boundary) suggest a range of solutions, but these solutions fail to achieve sufficient levels of entropy and produce too small segments, using the aggregate results is more reasonable.

Step 3: Run the PLS-POS Procedure

Running the PLS-POS procedure requires the researcher to make several choices regarding the algorithm settings. In the following, we explain the main steps of a PLS-POS analysis—as shown in Exhibit 6.5—and address the different choices to make. For a detailed explanation of the PLS-POS algorithm, see Becker, Rai, Ringle, and Völckner (2013).

When starting the PLS-POS algorithm, the first choice to make is the number of segments to retain from the data. As indicated earlier, researchers should draw on the FIMIX-PLS results along with practical considerations to make this decision. Using the segment number as input, PLS-POS then randomly assigns each observation to one of the segments and continues with the analysis. Alternatively, PLS-POS can use the final partition resulting from the FIMIX-PLS analysis as input. That is, each observation is assigned to the segment where it has the highest probability of membership as indicated by FIMIX-PLS.

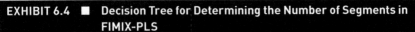

EXHIBIT 6.4 ■ Decision Tree for Determining the Number of Segments in FIMIX-PLS

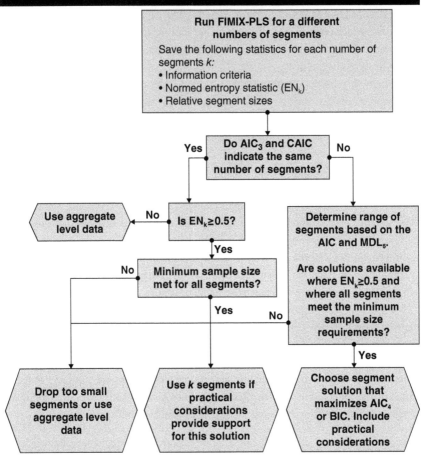

In PLS-POS's initialization stage, researchers have the option of running a **presegmentation**. If a presegmentation is chosen, PLS-POS uses the starting partition and then reassigns all observations at the same time to their closest segment in the first round—before reassigning observations one by one when running the subsequent iterations of the PLS-POS algorithm (Step 2 and Step 3). Reassigning all observations at the same time in the first round does not guarantee that all changes contribute to improving the optimization criterion. However, this option can save some time since it usually entails a significant improvement of the starting partition. The presegmentation process demonstrates its value only when PLS-POS initializes the starting partition randomly. In this case, researchers should run PLS-POS several times since the quality of the final solution depends on the starting partition. Similar to FIMIX-PLS, one should select the best solution—regarding the final value of the optimization criterion—of 10 repetitions to ensure convergence in a (close to the) global optimum

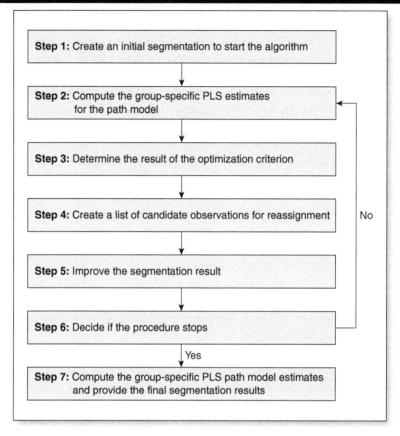

EXHIBIT 6.5 ■ The Main Steps of the PLS-POS Procedure

solution. However, when using the FIMIX-PLS solution as the starting partition, we recommend not using the presegmentation option as this step entails considerable changes in the FIMIX-PLS solution. As the FIMIX-PLS solution already converged in the global or a local optimum solution, the changes that come with the presegmentation would distort the FIMIX-PLS analysis.

Also, researchers need to decide on the **search depth**, which determines the maximum number of candidate observations for segment reassignment (see Step 4). If the search depth has a value of, for example, 10, PLS-POS analyzes only the first 10 candidates in the sorted list of observations for reassignment. Usually, researchers should analyze all candidate observations and, therefore, use a search depth equal to the number of observations. Nevertheless, reducing the search depth can be advantageous in complex segmentation settings (i.e., a large number of observations and/or a high number of segments) to reduce computation time. For the final PLS-POS computations, however, we recommend using a search depth equal to the number of observations in the full dataset.

Another decision to make in the initialization of the PLS-POS algorithm concerns the choice of the **optimization criterion** (Becker, Rai, Ringle, & Völckner, 2013). Consistent with the PLS-SEM algorithm's predictive nature (Lohmöller, 1989, Chapter 2; Wold, 1982, 1985), the objective of PLS-POS is to form groups of observations that maximize the model's in-sample predictive power (Hair, Risher, Sarstedt, & Ringle, 2019; Hair & Sarstedt, 2021; Sarstedt & Danks, 2022). For this purpose, a suitable optimization criterion is to maximize the sum of the endogenous constructs' explained variance (i.e., their R^2 values) across all the groups (clustering for maximum prediction; Anderberg, 1973). Alternatively, researchers can use the weighted sum of each group's sum of R^2 values as the optimization criterion, whereby the relative segment sizes of the different groups serve as the weights. As a consequence, the R^2 values of the larger segments contribute more strongly to the computation of the optimization criterion compared to smaller segments. Finally, researchers can focus the optimization criterion only on a particularly important endogenous construct in the PLS path model—similar to model selection tasks using information criteria (Danks, Sharma, & Sarstedt, 2020). Examples of such key target constructs are customer satisfaction and customer loyalty in customer satisfaction models (e.g., Fornell, Johnson, Anderson, Cha, & Bryant, 1996) and intention to use in technology acceptance models (e.g., Al-Gahtani, Hubona, & Wang, 2007). Nevertheless, we generally recommend considering all endogenous constructs, using the weighted sum of each group's sum of R^2 values as the optimization criterion.

In Step 2 and Step 3, PLS-POS estimates the group-specific PLS path models and uses the results as input for the computation of the optimization criterion. In Step 4, the method identifies a set of candidate observations for reassignment from their current segment to an alternative segment in order to improve the optimization criterion. Following Squillacciotti (2005, 2010), PLS-POS employs a distance measure to reassign observations. The closer the distance of an observation to a segment, the higher is the (in-sample) predictive power of this observation with respect to that segment. The distance measure draws on the structural model residuals, which are the differences between the estimated and actual scores of the endogenous construct(s). More precisely, the distance measure uses the observations' (in-sample) **prediction errors**, which are equal to the squared structural model residuals. Note that the residuals are squared so that larger differences between estimated and actual scores contribute more strongly to the prediction error than small differences. The sum of squared residuals (i.e., the prediction errors) represents the distance value.

While the actual endogenous construct scores result from the group-specific PLS-SEM analyses, the estimated scores need to be computed separately. This is done by multiplying the observations' exogenous construct scores with the corresponding path coefficient estimates and computing the sum of these products. To illustrate the computation of the estimated construct scores and the prediction error, consider the example of two equally sized segments and their segment-specific path coefficient estimates as shown in Exhibit 6.6. Furthermore, consider an observation belonging to Segment

EXHIBIT 6.6   ■   Example of the In-Sample Prediction Error Computation

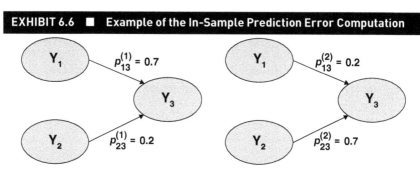

Segment 1   **Segment 2**

1 with (standardized) construct scores of 0.1 for construct Y_1, 0.6 for construct Y_2, and 0.5 for construct Y_3. To determine how well this observation estimates the target construct Y_3 in its own segment, PLS-POS multiplies the observation's construct score for Y_1 with the path coefficient $p_{13}^{(1)}$ (i.e., $0.1 \cdot 0.7 = 0.07$) as well as the observation's construct score for Y_2 with the path coefficient $p_{23}^{(1)}$ (i.e., $0.6 \cdot 0.2 = 0.12$). Adding these two products yields an estimated construct score for Y_3 of $0.07 + 0.12 = 0.19$ for this particular observation. The actual construct score of Y_3 for this observation is 0.5, yielding an in-sample prediction error (i.e., squared residual) of $(0.19 - 0.5)^2 = 0.0961$. Correspondingly, this observation's in-sample prediction error for Segment 2 is given by $[(0.1 \cdot 0.2 + 0.6 \cdot 0.7) - 0.5]^2 = 0.0036$. We find that the observation has a higher prediction error in its own Segment 1, compared to Segment 2. In other words, the observation has a higher in-sample predictive power in Segment 2 because the squared residual is smaller than the one in Segment 1. The sum of the squared residuals represents each observation's final distance values to its current segment and possible alternative segment assignments.

Based on the observations' distance values, PLS-POS generates a list of candidate observations for reassignment in Step 4 of the algorithm (Exhibit 6.5). If an observation has the shortest distance to its own segment, this observation does not become a candidate for reassignment. On the contrary, if an observation is closer to another than its own segment (as explained for the example shown in Exhibit 6.6), it fits better into that segment in terms of its in-sample predictive power, thereby potentially improving the optimization criterion after reassignment. If an observation has a shorter distance to more than one other segment, the segment with the shortest distance is considered in the list. Subsequently, the candidate observations are sorted in descending order of distances. If the list of candidates for reassignment is empty, the algorithm proceeds to Step 7 and stops with the final results' computation (Exhibit 6.5).

In Step 5, PLS-POS reassigns the first observation in the list of candidates to the alternative segment to which it has the shortest distance. A key challenge of this approach is the indeterminacy of the data assignment task as it is unknown how the group-specific PLS results will change after an observation is reassigned to a different group. For this reason, the PLS-POS algorithm exchanges only one observation at a time. For the changed data groups, the method updates the segment-specific PLS-SEM results and the optimization criterion. If the optimization criterion improves, the algorithm maintains the reassignment and proceeds with Step 6. Otherwise, if the optimization criterion does not improve, the algorithm does not reassign the observation. Then, the algorithm continues in the same way with the next candidate observation. As a result, the PLS-POS algorithm reassigns an observation only if this step improves the new segmentation solution's optimization criterion.

In Step 6, PLS-POS checks whether the maximum number of iterations has been reached. If not, the algorithm continues with the next iteration starting with Step 2. Otherwise, the algorithm continues with Step 7, the final step, which computes the PLS-POS results of the segmentation solution including the group-specific PLS path model estimates. The maximum number of iterations should be sufficiently high to obtain a solution that is close to the global optimum. We suggest using a maximum number of iterations equal to the higher number of the following options: 1,000 or two times the number of observations.

In summary, PLS-POS accounts for heterogeneity in the PLS path model by using a distance measure that facilitates the reassignment of observations in an effort to improve the (in-sample) prediction-oriented optimization criterion. The method is generally applicable to all PLS path models regardless of the mode of measurement model (for more details, see Becker, Rai, Ringle, & Völckner, 2013), the distribution of the data, or the complexity of the structural model. Like the EM algorithm in FIMIX-PLS, PLS-POS faces the problem that the algorithms could result in a local optimum solution due to its use of a hill-climbing approach. Hill climbing is a heuristic optimization method that belongs to the family of local search procedures. The method starts with a random partition into a prespecified number of segments and iteratively improves this solution one element at a time until it arrives at a (locally) optimized solution. Thus, if researchers do not build on the FIMIX-PLS starting partition, a repeated application of PLS-POS (e.g., 10 times per prespecified number of segments) with different starting partitions is advisable.

Step 4: Explain the Latent Segment Structure

Upon convergence, PLS-POS provides users with each observation's segment membership along with the segment-specific model estimates for the measurement and structural models. As the identified segments are—by definition—latent, the model estimates do not offer any indication of how the segments were formed. Turning the initial results into actionable understanding requires the researcher to interpret the segments in terms of observable and managerially meaningful variables. To do so, we need to translate the latent segment structure into an observable one by partitioning the data using one or more explanatory variable(s). The resulting partition of the data should largely correspond to the grouping produced by PLS-POS. Such an analysis is also referred to as **ex post analysis**.

An ex post analysis could involve running a binary (in case of two latent segments) or multinomial (in case of three or more latent segments) logistic regression with the segment affiliation as the dependent variable and potential explanatory variables as the independent variables (e.g., Money, Hillenbrand, Henseler, & Da Camara, 2012; Wilden & Gudergan, 2015). Comparing the resulting log odds allows identifying those independent variables that have the strongest effect on the latent segment structure. Using cross tabs, researchers can then contrast the frequency distribution of the PLS-POS segment affiliation variable with one or more explanatory variables (e.g., Dessart, Aldás-Manzano, & Veloutsou, 2019).

Exhibit 6.7 shows a cross tab example for 400 observations, two PLS-POS segments, and the explanatory variable *country of origin*. The high frequencies indicate that the majority of respondents in PLS-POS Segment 1 are from the United States. On the contrary, PLS-POS Segment 2 primarily includes European respondents. More precisely, the variable *country of origin* explains 188 + 177 = 365 of 400 PLS-POS segment assignments, which is an overlap of 365 / 400 = 91.25%. If each cell included about 100 respondents, *country of origin* would be unsuitable for explaining the PLS-POS segmentation and the overlap with the explanatory variable would not be systematic. In general, the more equal the distribution across the cells, the less useful is the explanatory variable, and vice versa.

Researchers have also drawn on tree-building algorithms such as the Chi-squared automatic interaction detector (CHAID) or classification and regression trees (CART) to identify suitable explanatory variables (e.g., Ringle, Sarstedt, & Mooi, 2010; Sarstedt & Ringle, 2010). These techniques can be used to predict values of a categorical variable (in our case, the observations' latent segment affiliations) from one or more continuous or categorical predictor variables. Both CHAID and CART techniques construct trees, in which each (nonterminal) node identifies a split condition, to yield an optimum classification. The resulting nodes can then be used to identify combinations of explanatory variables that exhibit a large overlap with the latent class segmentation.

EXHIBIT 6.7 ■ Cross Tab Example

	Country of Origin		
	United States	**Europe**	*Sum*
PLS-POS Segment 1	188 (47.00%)	14 (3.50%)	**202 (50.50%)**
PLS-POS Segment 2	21 (5.25%)	177 (44.25%)	**198 (49.50%)**
Sum	**209 (52.25%)**	**191 (47.75%)**	**400 (100.00%)**

Importantly, to successfully run an ex post analysis, researchers must be able to consider a wide range of observable characteristics that can serve as explanatory variables. Examining too few observable characteristics restricts the researcher's ability to reproduce the PLS-POS partition. With this in mind, researchers should assess whether a single explanatory variable, or a set of variables, has theoretical meaning to elucidate possible differences in path coefficients across identified segments. Therefore, assessing the explanatory role of possible variables, so that the latent class analysis can be implemented more completely, should be considered in the research design stage when collecting descriptive or other variables that can matter.

Nevertheless, reproducing the PLS-POS partition remains a challenging task, as observable characteristics often do not match the latent segment structures well. Against this background, an overlap of 60% between the PLS-POS partition and the one produced by the explanatory variable(s) can be considered satisfactory.

Step 5: Analyze group-specific models

Once the researcher has identified one or more explanatory variables that match the PLS-POS partition well, the final step is to estimate group-specific models as indicated by the explanatory variable(s). In doing so, the researcher must ensure that all the model measures meet common quality standards as documented in, for example, Hair, Hult, Ringle, and Sarstedt (2022). These analyses complete the latent class analysis. However, further analyses can involve testing whether the differences between segment-specific path coefficients are also significantly different by means of multigroup analysis, which must also include testing for measurement invariance as described in Chapter 5.

Exhibit 6.8 summarizes the rules of thumb for a latent class analysis in PLS-SEM. The use of FIMIX-PLS for determining the number of segments requires particular care. Sarstedt, Radomir, Moisescu, and Ringle (2022) reviewed 45 applications of FIMIX-PLS in major business research journals and found that researchers frequently disregard important steps in their analyses or misapply them. For example, determining the number of segments frequently occurs ad hoc, without clear reference to information criteria or relative segment sizes. Another point of concern relates to researchers' processing of the FIMIX-PLS results. Software programs such as SmartPLS provide segment-specific path coefficients, which result from weighted regressions using the relative segment sizes as input. As a result, these parameter estimates remain abstract because they do not refer to a unique partition, as is the case with hard clustering (i.e., the full and thus unique assignment of an observation to a particular segment). Instead, researchers should assign the observations to the segments based on their highest probability of segment membership. Finally, only a few studies report an ex post analysis, despite its obvious relevance for deriving managerial recommendations. A likely reason for this finding is the lack of explanatory variables that allow for reproducing the latent partition produced by FIMIX-PLS.

EXHIBIT 6.8 ■ Rules of Thumb for a Latent Class Analysis

FIMIX-PLS algorithm settings:

● Use a stop criterion 10^{-7} of and a maximum number of 5,000 iterations.

● Use the best solution of at least 10 repetitions to avoid convergence in a local optimum.

● When the indicators have missing values, use casewise deletion. Do not use mean value replacement or any other imputation method.

● To define a range of reasonable segment numbers, use one segment as the lower bound and the largest integer when dividing the sample size by the minimum segment sample size as the upper bound.

Determining the number of segments to retain:

● Information criteria: If AIC_3 and CAIC indicate the same number of segments, choose this solution. Also consider the segment number as indicated by AIC_4 and BIC. Generally, choose fewer segments than indicated by AIC and more segments than indicated by MDL_5.

● The EN should be higher than 0.5.

● Ensure that the segment-specific sample sizes meet the minimum sample size requirements. If these are not met, reduce the number of segments, or discard the extraneous segments and focus on the remaining larger ones.

● Take practical considerations into account. If possible, let a priori information and theory contribute to substantiating segments. Ensure that the solution is relevant in managerial terms.

PLS-POS algorithm settings:

● Use the FIMIX-PLS results to determine the number of segments to retain from the data. Consider taking the FIMIX-PLS solution as the starting partition for PLS-POS.

● Do not use the presegmentation option when using the FIMIX-PLS solution as the starting partition for PLS-POS.

● When using random assignment as the starting partition, use the best solution of 10 repetitions to avoid convergence in a local optimum.

● Select a high maximum number of iterations that the PLS-POS algorithm will perform. This number should equal the higher one of the following options: 1,000 or two times the number of observations.

● The search depth to identify candidate observations for segment reassignment should equal the number of observations.

● Consider using the weighted sum of all constructs' R^2 values as the optimization criterion.

EXHIBIT 6.8 ■ Rules of Thumb for a Latent Class Analysis *(Continued)*

Ex post analysis:

● Partition the data by using an explanatory variable, or a combination of several explanatory variables, which yields a grouping of data that largely corresponds to the one produced by PLS-POS.

● The ex post analysis needs to be considered in the data collection stage. Researchers are advised to gather information on as many explanatory variables as reasonable.

● A 60% overlap between the PLS-POS partition and the one produced by the explanatory variable(s) is regarded as satisfactory. Consider running a PLS multigroup analysis after establishing (partial) measurement invariance.

CASE STUDY ILLUSTRATION

Step 1: Run the FIMIX-PLS Procedure

Our illustration of the latent class analysis is partially taken from Matthews, Sarstedt, Hair, and Ringle's (2016) case study on the use of FIMIX-PLS, which also draws on the extended corporate reputation model. In light of the problems that could arise when using mean imputation to handle missing data, we deleted all observations with missing values in the data examination stage (i.e., casewise deletion) prior to the actual analysis. After casewise deletion, 336 observations remain. Navigate to the SmartPLS **Workspace** window on the left-hand side. The reduced dataset with the name **Complete data after casewise deletion** is already included in the SmartPLS project labeled **Example – Corporate reputation (advanced)**, which we use for all case study illustrations in this book (see Chapter 1). The sample project is also available at https://www.pls-sem.com. Double-click on **Corporate reputation model** to open the extended PLS path model for the corporate reputation example in the SmartPLS modeling window. Above the list of indicators, SmartPLS shows the dataset that will be used for the model estimation. Click on the **Select dataset** button and choose the **Complete data after casewise deletion** dataset. Then estimate the model by using **Calculate → PLS-SEM algorithm** in the SmartPLS menu with the default settings. Note that we obtain the same results when using the full dataset with 344 observations and selecting **Casewise deletion** in the **Missing values** tab in the starting dialog when running the FIMIX-PLS method. However, for illustration purposes, we use the **Corporate reputation data after casewise deletion** with 336 observations.

To initiate the FIMIX-PLS analysis, click on **Calculate → Finite mixture (FIMIX) segmentation** in the menu. Alternatively, we can select this option after a left-click on the **Calculate** symbol in the toolbar. After selecting the **Finite mixture (FIMIX) segmentation** option, the dialog box as displayed in Exhibit 6.9 appears. For the initial analysis, start with a one-segment solution (i.e., enter a **1** in the box next to **Number of segments**) and use the default settings for the **Maximum iterations** number (i.e., **5,000**), the **Stop criterion** (i.e., **10⁻⁷**), the **Number of repetitions** (i.e., **10**), and the **Random number**

EXHIBIT 6.9 ■ FIMIX-PLS Dialog Box

generator (i.e., **Fixed seed**). Also make sure to select **Open report** at the bottom of the dialog box. The dialog box has another tab to specify the standard PLS-SEM algorithm settings. Use the default setting for the PLS-SEM algorithm. The tab for missing value treatment is not available as the **Complete data after casewise deletion** has no missing values. Finally, click on **Start calculation**.

After convergence, SmartPLS opens the **Results report**. Before analyzing the results in detail, we need to rerun FIMIX-PLS for higher-segment solutions (e.g., two to five segments). To determine the upper bound of the range of segment solutions, check the minimum sample size requirements as specified in Hair, Hult, Ringle, and Sarstedt (2022). With a maximum number of eight arrowheads pointing at any construct in the model (formative indicators of $QUAL$) and assuming a 5% significance level, as well as a minimum R^2 of 0.25, we would need 54 observations to reliably estimate the model. The greatest integer from dividing the sample size (i.e., 336) by the minimum sample size (i.e., 54) yields a theoretical upper bound of six segments (Sarstedt, Ringle, & Hair, 2017). However, as an equal distribution of observations—which would be necessary to meet the minimum sample size requirements—is highly unlikely, assuming a smaller maximum number of segments seems reasonable (e.g., five segments). Note that following the more rigorous inverse square root method (Kock & Hadaya, 2018) and assuming a power level of 80%, a significance level of 5%, and a minimum path coefficient of 0.20 would require each extracted segment to have at least 155 observations (Hair, Hult, Ringle, & Sarstedt, 2022, Chapter 1). Hence, extracting more than two segments is not feasible when applying the inverse square root method on the current dataset. For illustrative purposes, however, we apply the standard power table-based minimum sample size recommendations, considering a maximum number of five segments. We therefore rerun the procedure four times, specifying **2** to **5** under **Number of segments** in the FIMIX-PLS dialog box with otherwise the same settings as applied before. SmartPLS automatically retains the results. SmartPLS also allows saving the results of each FIMIX-PLS run (as well as all other analyses). To do so, simply click on the **Save** button in the toolbar of the **Results report**. After selecting the project and a name, SmartPLS will store the report as a separate project file in the **Workspace** window.

Step 2: Determine the Number of Segments

To determine the number of segments to retain from the data, we need to examine the fit indices, which we can find under **Quality criteria → Model selection criteria** in each of the FIMIX-PLS results reports. To facilitate their comparison across the different segment number solutions, we can export the values to a spreadsheet program, such as Microsoft Excel. To do so, click on **Copy to Excel** in the **Model selection criteria** window or on **Excel** in the toolbar, and paste the values into an Excel file. Exhibit 6.10

EXHIBIT 6.10 ■ Model Selection Criteria for a One- to Five-Segment Solution					
	Number of Segments				
Criteria	*1*	*2*	*3*	*4*	*5*
AIC	2,847.72	2,777.658	2,752.377	2,744.696	**2,707.552**
AIC_3	2,864.72	2,812.658	2,805.377	2,815.696	**2,796.552**
AIC_4	2,881.72	**2,847.658**	2,858.377	2,886.696	2,885.552
BIC	2,912.61	**2,911.257**	2,954.684	3,015.711	3,047.275
CAIC	**2,929.61**	2,946.257	3,007.684	3,086.711	3,136.275
MDL_5	**3,308.18**	3,725.653	4,187.912	4,667.771	5,118.167
LnL	−1,406.86	−1,353.829	−1,323.189	−1,301.348	**−1,264.776**
EN	Not available	0.430	0.589	0.538	**0.718**

Note: Numbers in bold represent the best outcome per criterion (i.e., for a comparison across five segments).

provides an overview of the log-likelihood (LnL) values, information criteria, and the EN for a one- to five-segment solution.

Two aspects are worth noting: First, for each model selection criterion, the optimal solution is the number of segments with the lowest value (see bold numbers in Exhibit 6.10), except in terms of EN, where higher values indicate a better separation of the segments. Second, our values will be the same as those shown in Exhibit 6.10, only if we also selected the **Fixed seeds** option in the algorithm settings. If we selected the **Random seeds** options, SmartPLS used different random values in every FIMIX-PLS iteration. However, in this case, the conclusions regarding the number of segments to retain should not differ from those presented here.

Interpreting the results, we find that AIC_3 and CAIC do not indicate the same number of segments, and neither do AIC_3 and BIC. AIC shows a five-segment solution, suggesting that the correct number likely is lower than this. On the other extreme, CAIC and particularly MDL_5 result in a one-segment solution, suggesting that more than one segment should be considered. Further analysis reveals that two other criteria with good performance in detecting an appropriate number of segments, AIC_4 and BIC, both indicate two segments, thus providing initial support for this solution. However, the two-segment solution exhibits an EN value below 0.50, suggesting that the two segments are not well separated.

In the results **Report**, under **Final results → Segment sizes**, SmartPLS shows the relative segment sizes of the FIMIX-PLS solution for a certain prespecified number of segments. Examining the relative segment sizes across FIMIX-PLS solutions for different numbers of segments, as summarized in Exhibit 6.11, shows that selecting more than two segments is not reasonable. Note that these segment sizes do not result from a hard clustering of observations based on the maximum probabilities of segment membership but from weighted least squares regressions using these probabilities as input. For example, for a three-segment solution, the breakdown of segment sizes is Segment 1 with 45.3% (approx. 153 of 336 observations), Segment 2 with 44.3% (approx. 148 of 336 observations), and Segment 3 with only 10.4% (approx. 35 of 336 observations). As can be seen, with 35 observations,

EXHIBIT 6.11 ■ Relative Segment Sizes

Number of Segments	Relative Segment Sizes				
	Segment 1	Segment 2	Segment 3	Segment 4	Segment 5
2	0.515	0.485			
3	0.453	0.443	0.104		
4	0.326	0.262	0.224	0.188	
5	0.532	0.312	0.088	0.045	0.023

Segment 3 is too small for a segment-specific PLS-SEM analysis. One way to handle this is to consider dropping the third segment and instead focus on the analysis and interpretation of the other two larger segments. However, in light of the results we continue with our interpretation of a two-segment solution.

Open the SmartPLS **Results report** for two segments and go to **Quality criteria → R-square** to view the R^2 values for the two segments and the aggregate dataset. Exhibit 6.12 shows the segment-specific R^2 values along with the weighted R^2 values that result from multiplying the segment-specific values with the relative segment sizes. The R^2 values in Segment 1 are considerably higher compared to the full set of data, while those in Segment 2 are lower. We also find that the weighted average R^2 values of the FIMIX-PLS two-segment solution are only slightly higher than those of the full set of data. These results substantiate our reservations regarding the FIMIX-PLS segmentation solution. The sample seems to be relatively homogeneous. However, when inspecting the FIMIX-PLS segment-specific path coefficients in the **Results report** under **Final results → Path coefficients**, we find that some differences in the segment-specific path coefficient estimates. For example, the path from *CUSA* to *CUSL* differs by more than 0.15 points across the two segments.

Overall, these results suggest that our model estimates using the aggregate dataset are not substantially impacted by heterogeneity. The fit indices yield ambiguous results regarding the number of segments to retain from the data while the relative segment sizes permit only a two-segment solution. The EN criterion, however, indicates that the two-segment solution produces fuzzy results in terms of the observations' probabilities of segment membership. Finally, the (overall) R^2 of the final target construct *CUSL* is only slightly higher when considering the FIMIX-PLS results ($R^2 = 0.604$) rather than the aggregate results ($R^2 = 0.562$). Hence, extracting two segments or more for further analyses does not appear to be beneficial. However, for illustrative purposes and to outline how PLS-POS can be applied, we take a more detailed look at the two-segment solution using PLS-POS.

Step 3: Run the PLS-POS Procedure

In the next step, we run the PLS-POS procedure by clicking on **Calculate → Prediction-oriented segmentation (POS)** in the menu bar. Alternatively, we can left-click on the wheel symbol in the toolbar and select the corresponding option in

EXHIBIT 6.12 ■ **FIMIX-PLS R^2 Values for the Two-Segment Solution**

R-square   Zoom (100%)   Copy to Excel   Copy to R

	Original sample R-squares	Weighted average R-squares	Segment 1	Segment 2
COMP	0.629	0.670	0.812	0.519
CUSA	0.293	0.353	0.530	0.164
CUSL	0.562	0.604	0.807	0.389
LIKE	0.557	0.594	0.814	0.361

the combo box that opens. In the dialog box that opens (Exhibit 6.13) select **2** under **Groups**—as indicated by the previous FIMIX-PLS analysis. The dataset has 336 observations after casewise deletion. Since multiplying 336 observations by 2 yields the number of 672, which is lower than 1,000, select the default value of **1,000** for the **Maximum iterations**. The **Search depth** should equal the number of observations, which is 336. If the default value of **1,000** remains unchanged, SmartPLS automatically stops when 336 has been reached since it is not possible to search beyond the maximum number of observations in the dataset. Moreover, we select the **FIMIX segmentation** instead of the **Random assignment** option for the **Initial separation** in PLS-POS and do not opt for applying the **Pre-segmentation** option. Furthermore, we do not want to focus on a single endogenous construct in the corporate reputation model such as *CUSL*. Instead, we want to maximize the group-specific explanation of all endogenous constructs at the same time. We therefore select the **Sum of all constructs weighted R-squares** as the **Optimization criterion**. Finally, we use **Fixed seed** option for the **Random number generator**. Check the **Open report** box and initiate the analysis by clicking on **Start calculation**.

EXHIBIT 6.13 ■ Smart PLS PLS-POS Dialog Box

POS setup	PLS setup	FIMIX setup

Groups	2
Maximum iterations	1000
Search depth	1000
Initial separation	FIMIX segmentation
	☐ Pre-segmentation
Optimization criterion	Sum of all constructs weighted R-squares
Target construct	
Random number generator	Fixed seed

✓ Open report

Default settings   Close   Start calculation

After convergence, SmartPLS opens the **Results report**. We can save this report for later use in our SmartPLS project by clicking **Save** on the toolbar. To avoid premature convergence in local optima, we should run PLS-POS multiple times and select the solution with the highest **Objective criterion** value, which we find in the PLS-POS **Results report** under **Quality criteria → Change in objective criterion**. We suggest running the procedure **10** times, as this number achieves a good balance between running time and results quality (i.e., in providing a near to the global optimum solution). Hence, rerun PLS-POS 10 times, save each obtained solution, and finally select the one with the highest **Objective criterion**.

Having identified the best solution, we continue the results analysis by examining the R^2 values under **Quality Criteria → R-square**. In addition, we analyze the (relative) segment sizes, which we can find under **Final results → Segment sizes**. Exhibit 6.14 summarizes the PLS-POS results of the two-segment solution. While Segment 1 (relative segment size: 56.8%) returns substantially higher R^2 values, the results for Segment 2 (relative segment size: 43.2%) are at similar or higher levels compared to the outcomes of the full dataset. Importantly, all average weighted R^2 values produced by PLS-POS are higher than the R^2 values when analyzing the aggregate dataset. This particularly holds for the model's final target construct *CUSL*.

Exhibit 6.15 shows the segment-specific path coefficients from the PLS-POS analysis. We find substantial differences in the path coefficient estimates across the two groups. For example, while *COMP* has a pronounced positive impact on *CUSL* in the first segment (0.242), the same effect is negative in the second segment (–0.302). However, in the full dataset, the corresponding path coefficient is close to zero (0.011).

Step 4: Explain the Latent Segment Structure

To interpret the segments in terms of observable and practically meaningful variables, we next run an ex post analysis to identify one or more explanatory variable(s) that match the PLS-POS partition in the best possible way. To begin with, we first have to transfer the partition as indicated by PLS-POS to the original dataset. Select the option **Create data file** in the **Results report** of PLS-POS. In the box that opens, choose an intuitive name (e.g., **Corporate reputation data 336 PLS-POS**) under **File name**, check all boxes below **Select columns**, and click the **Create** button. Return to the Workspace window in SmartPLS (i.e., click on the arrow button with the label

EXHIBIT 6.14 ■ PLS-POS Segmentation Solution

R-square   Zoom (100%)   Copy to Excel   Copy to R

	Original sample R-squares	Weighted average R-squares	POS segment 1	POS segment 2
COMP	0.629	0.704	0.770	0.618
CUSA	0.293	0.478	0.521	0.421
CUSL	0.562	0.641	0.706	0.555
LIKE	0.557	0.642	0.631	0.657

EXHIBIT 6.15 ■ Path Coefficients of the Full Set of Data and the Two PLS-POS Segments				
	Full Dataset	*Segment 1*	*Segment 2*	*Segment 1– Segment 2**
ATTR → COMP	0.085	0.206	−0.115	0.321
ATTR → LIKE	0.159	0.144	0.113	0.031
COMP → CUSA	0.135	0.673	−0.438	1.112
COMP → CUSL	0.011	0.242	−0.302	0.544
CSOR → COMP	0.057	−0.129	0.194	−0.323
CSOR → LIKE	0.190	0.267	0.168	0.099
CUSA → CUSL	0.510	0.586	0.181	0.405
LIKE → CUSA	0.445	0.069	0.808	−0.739
LIKE → CUSL	0.334	0.096	0.747	−0.651
PERF → COMP	0.295	0.397	0.175	0.223
PERF → LIKE	0.116	0.267	−0.114	0.381
QUAL → COMP	0.431	0.455	0.588	−0.133
QUAL → LIKE	0.378	0.223	0.692	−0.470

* Instead of the frequently used absolute difference, we use the Segment 1 – Segment 2 difference of the segments' specific path coefficients. As a result, a positive sign of the difference indicates that the path coefficient of Segment 1 is larger than the one of Segment 2, and vice versa.

Edit in the left upper corner to return to the modeling window, followed by the arrow button with the label **Back** to return to the Workspace window). In the **Example – Corporate reputation (advanced) project**, we now see newly generated dataset with the name **Corporate reputation data 336 PLS-POS**. Double-click on this dataset to open the **Data view** in SmartPLS and choose the **Export to EXCEL / CSV** option in the toolbar. Save the dataset (e.g., by using the file name **Corporate reputation data 336 PLS-POS**).

We can now use this dataset to compare the PLS-POS partition with those indicated by other observable variables in the dataset. Unfortunately, the corporate reputation dataset has only two such variables, which indicate each respondent's service provider (service provider 1 – 4) and the type of service the respondent uses (prepaid or contract). Therefore, the chances of reproducing the PLS-POS partition adequately are relatively low. We strongly recommend including as many potential explanatory variables as reasonable in the survey development to increase the chances of adequately reproducing the latent segments in the ex post analysis.

As an illustration, in this application we use simple crosstabs to compare the PLS-POS partition with the one produced by the variables *serviceprovider* and *servicetype*. For this purpose, we employ a statistics program such as IBM SPSS Statistics or JASP, open the extended data file (e.g., **Corporate reputation data 336 pls-pos.csv**), and use the crosstabs analysis option for the *serviceprovider* and PLS-POS groups variables (see Chapter 5 in Sarstedt & Mooi, 2019, for an illustration of cross tabs using IBM SPSS Statistics). Exhibit 6.16 shows the results of this analysis.

Comparing the cell counts, we find that the best match is achieved when assigning respondents who use service provider 1, 2, or 3 to PLS-POS Segment 1, and those who use service provider 4 to PLS-POS Segment 2. Using this grouping (81 + 71 + 22 + 25) / 336 = 59.23% of the respondents match the PLS-POS partition. However, matching service provider 4 with PLS-POS Segment 2 produces a segment with only 42 observations. Since a sample of 42 observations is not sufficient to reliably estimate a PLS path model for Segment 2, we also assign the users of service provider 3 to Segment 2 because service provider 3 has almost the same number of respondents belonging to PLS-POS Segments 1 and 2. Therefore, we consider respondents who use service providers 1 or 2 as matching the first latent segment, whereas we think of users of service providers 3 and 4 as matching the second latent segment. This grouping is also reasonable from a content point of view, as service providers 1 and 2 represent two large multinational telecommunication providers while service providers 3 and 4 are smaller telecommunication providers that operate in only a few countries.

Using this grouping, (81 + 71 + 21 + 25) / 336 = 58.93% of the respondents match the PLS-POS partition. Even though the overlap is only slightly below the cutoff value of 60%, this result is not very satisfactory. In addition, whereas FIMIX-PLS and PLS-POS provided almost equally sized groups, the segments retained through the ex post analysis are highly unequal in their size. Alternatively, when contrasting the PLS-POS partition with the one produced by the *servicetype* variable (see Chapter 5), the results are even worse, with an overlap of merely 52.68%. Since no further explanatory variables are available in the dataset, we continue the illustration by using *serviceprovider* as an explanatory variable. Note that clear descriptions of identified segments are rather challenging in general.

EXHIBIT 6.16 ■ Cross Tab of PLS-POS Partition and Service Provider			
	PLS-POS Segments		
Service Provider	*1*	*2*	*Sum*
1	81 (24.11%)	47 (13.99%)	**128 (38.10%)**
2	71 (21.13%)	52 (15.48%)	**123 (36.61%)**
3	22 (6.55%)	21 (6.25%)	**43 (12.80%)**
4	17 (5.06%)	25 (7.44%)	**42 (12.50%)**
Sum	**191 (56.85%)**	**145 (43.15%)**	**336**

Step 5: Analyze group-specific models

The next step in the ex post analysis is to compute the segment-specific results. For this purpose, we need to define *serviceprovider* as a grouping variable. However, *serviceprovider* has four unique values, which we need to condense into two groups. In our case, the service provider values 1 and 2 correspond to the first PLS-POS segment, while the values 3 and 4 correspond to the second PLS-POS segment. Double-click on the **Complete data after casewise deletion [366]** dataset in the **Example – Corporate reputation (advanced) project**, which we find in the SmartPLS **Workspace**, and select the **Add group** button in the menu bar to define these two groups. Clicking on the **Add group** button opens a dialog box in which we can define new grouping variables (Exhibit 6.17). Under **Group name**, specify the new grouping variable's name (e.g., *serviceprovider 1 + 2*) and define the group selection options as shown in Exhibit 6.17 (Panel A). After clicking on **Apply**, SmartPLS will create a group that includes those observations for which the service provider is less than 3 (i.e., 1 and 2). In a similar manner, we create a new grouping variable (e.g., *serviceprovider 3 + 4*) with all observations for which the service provider value is higher than 2 (i.e., 3 and 4), as shown in Exhibit 6.17 (Panel B).

Having defined the groups, double-click on **Corporate reputation model** in the **Example – Corporate reputation (advanced) project**, which we can find in the

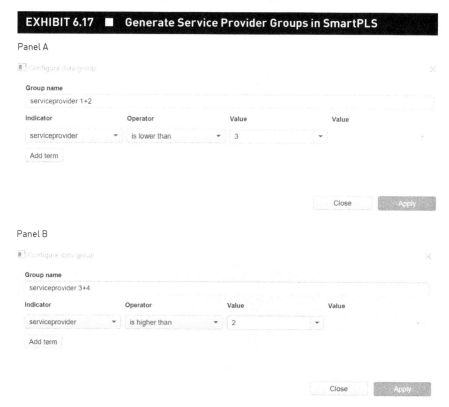

EXHIBIT 6.17 ■ Generate Service Provider Groups in SmartPLS

Panel A

Panel B

SmartPLS **Workspace**, and run the PLS-SEM algorithm by selecting **Calculate →
PLS-SEM algorithm** in the toolbar. The dialog box that opens offers the standard
options for running the PLS-SEM algorithm. Retain the default settings for the PLS-
SEM algorithm, the missing values, and the weighting, and click on the **Data** tab. As
shown in Exhibit 6.18, a box will open, from which we can select one or more groups to
be analyzed separately. Tick the boxes next to *serviceprovider 1 + 2* and *serviceprovider 3
+ 4* to select these two groups for the group-specific PLS path model estimations, or click
on **Group data sets** to select/deselect both groups simultaneously. Finally, check the box
next to **Open report** and start the analysis by clicking on **Start calculation**.

SmartPLS initially shows the results of the aggregate data analysis, but we can
easily scan the other groups by clicking on the pull-down menu below **Data group**
on the left-hand side of the screen. Use the up and down arrows on our keyboard
to switch between the groups and their displayed results in the modeling window.
We also need to separately run bootstrapping for each group by going to **Calculate
→ Bootstrapping**. Use the default settings, but select **10,000** subsamples, the
Complete (slower) option to get the HTMT bootstrap results, and the **One tailed**
test for the significance level of **0.05**. In the **Data** tab, we make sure that we select

EXHIBIT 6.18 ■ Data Groups in the PLS-SEM Algorithm

all the groups before clicking on the **Start calculation** button. Exhibit 6.19 provides an overview of the aggregate data and the group-specific results, including all the reflective and formative measurement model evaluation criteria as documented in, for example, Hair, Hult, Ringle, and Sarstedt (2022). The measurement model evaluation results support the measures' reliability and validity.

Comparing the path coefficient estimates of the two service provider groups shows clear differences in the structural model effects. For example, whereas the effect of *COMP* on *CUSL* is not significant in the first service provider group (i.e., the large telecommunication providers), it is significantly negative in the second group (i.e., the small telecommunication providers). This counterintuitive result casts doubt on the

EXHIBIT 6.19 ■ Final Service Provider Segmentation Results			
	Original Sample	*Service Provider 1 + 2*	*Service Provider 3 + 4*
N	336	251	85
Relative segment size	100.0%	74.7%	25.3%
	Path	Path	Path
ATTR → COMP	0.085	0.047	0.275***
ATTR → LIKE	0.159**	0.176**	0.001
COMP → CUSA	0.135**	0.143*	0.191
COMP → CUSL	0.011	0.097	−0.128*
CSOR → COMP	0.057	0.011	0.215**
CSOR → LIKE	0.190***	0.192***	0.105
CUSA → CUSL	0.510***	0.461***	0.611***
LIKE → CUSA	0.445***	0.501***	0.235**
LIKE → CUSL	0.334***	0.300***	0.414***
PERF → COMP	0.295***	0.357***	0.162
PERF → LIKE	0.116*	0.129	−0.017
QUAL → COMP	0.431***	0.454***	0.281**
QUAL → LIKE	0.378***	0.360***	0.664***
Reflective measures			
Convergent validity (AVE)	+	+	+
Reliability (composite reliability, Cronbach's α)	+	+	+

EXHIBIT 6.19 ■ Final Service Provider Segmentation Results *(Continued)*

	Original Sample	Service Provider 1 + 2	Service Provider 3 + 4
Discriminant validity (HTMT$_{inference}$)	+	+	+
Formative measures			
Convergent validity	+	+	+
Collinearity	+	+	+
Significance and relevance of the indicators	+	+	+
R^2 values			
COMP	0.629	0.646	0.654
CUSA	0.293	0.365	0.141
CUSL	0.562	0.557	0.625
LIKE	0.557	0.570	0.525
Weighted R^2 values			
COMP	0.629	0.648	
CUSA	0.293	0.308	
CUSL	0.562	0.574	
LIKE	0.557	0.559	

*** $p \leq 0.01$; ** $p \leq 0.05$; * $p \leq 0.10$.

+/– = measurement model evaluation criterion fulfilled/not fulfilled in accordance with Hair, Hult, Ringle, and Sarstedt (2022).

validity of the ex post analysis. Similarly, the antecedents' impact on *COMP* and *LIKE* varies substantially between the groups. For example, whereas in the first group, *CSOR* has a significant effect on *LIKE*, but not on *COMP*, the opposite holds for the second group. Further analyses should involve testing whether these differences in path coefficients are significant using multigroup analysis (Chapter 5).

Comparing the path coefficients from the two service provider groups with those from the PLS-POS analysis shows that the results do not align well. For example, the significant but relatively low relationship between *COMP* and *CUSA* (0.143) in service provider Segment 1 is considerably higher in the PLS-POS (0.673) solution. The same holds for most of the other structural model relationships, which suggest that the PLS-POS results cannot be adequately reproduced by using the service provider variable. This result is not surprising, given the limited overlap between the PLS-POS and service provider partitions.

To further assess the segmentation solution, we compute the weighted R^2 values as the sum of the segment-specific R^2 values, weighted by the relative segment size. For example, for *COMP* in the service provider grouping, the weighted R^2 is $0.747 \cdot 0.646 + 0.253 \cdot 0.654 = 0.648$. Next, we can compare the resulting weighted R^2 values (produced by the data grouping) with the overall R^2 values resulting from the aggregate data level analysis. This analysis shows that the grouping using the service provider variables increases the model's explanatory power compared to the aggregate level analysis.

Further analysis would involve running the MICOM procedure to assess whether the two groups establish measurement invariance (Chapter 5). If at minimum partial measurement invariance is established, we would continue the analysis by running multigroup analysis to evaluate whether the path coefficients differ significantly between the groups of large and small service providers.

SUMMARY

- **Comprehend the FIMIX-PLS and PLS-POS approaches for identifying and treating unobserved heterogeneity.** FIMIX-PLS is the first and most prominent latent class approach in PLS-SEM. Drawing on the mixture regression concept, FIMIX-PLS simultaneously estimates the path coefficients of all model relations for a predefined number of groups. FIMIX-PLS proves particularly useful for identifying the number of segments to extract from the data but does not reliably identify the underlying segment structure and disregards heterogeneity in the measurement models. PLS-POS, a hill-climbing approach that gradually reallocates observations between the segments with the aim of maximizing the overall explained variance across all segments, overcomes these limitations. The approach has been shown to reliably identify segment-specific path coefficients but does not give a clear indication regarding the number of segments to consider in the analysis.

- **Understand how to use FIMIX-PLS and PLS-POS jointly in a latent class analysis.** Considering each method's strengths and weaknesses, a latent class analysis in PLS-SEM first draws on FIMIX-PLS to identify the appropriate number of segments to retain from the data. The FIMIX-PLS solution then serves as a starting point for running PLS-POS. PLS-POS improves the FIMIX-PLS solution and allows considering heterogeneity in the measurement models of constructs. A subsequent ex post analysis aims at identifying explanatory variables that can be used to partition the data into (observable) groups, which largely correspond to the latent segment structure produced by PLS-POS. The final step involves testing segment-specific path models drawing on the segments that are defined by the previously identified explanatory variables.

- **Appreciate how to execute a FIMIX-PLS analysis using SmartPLS.** An important consideration before running FIMIX-PLS involves the treatment of missing values. In light of the considerable biases that would occur with this type of analysis, do not use mean value replacement for missing values. Instead, delete all observations with missing values by using casewise deletion. Running the FIMIX-PLS procedure itself requires making several choices regarding the algorithm settings. When running FIMIX-PLS, use a stop criterion of 10^{-7}, a maximum number of 5,000 iterations, and at least 10 repetitions to avoid convergence in a local optimum. To define a range of reasonable segment numbers, use one segment as the lower bound and the largest integer when dividing the sample size by the minimum segment sample size as the upper bound. This upper bound can be adjusted, depending on the complexity of the model and the sample size. The decision of how many segments to retain from the data should consider a combination of different information criteria, the EN statistic, and the relative segment sizes.

- **Know how to use SmartPLS to run a PLS-POS analysis.** Running PLS-POS in SmartPLS requires making several decisions. Researchers need to determine the number of segments to retain from the data. This decision should be based on the FIMIX-PLS results, either by specifying the segment number in line with the information and classification criteria or by using the FIMIX-PLS solution as the starting partition. When using a random initialization, researchers should run the PLS-POS algorithm 10 times to avoid convergence in local optima. In a random initialization process, researchers have the option of running a presegmentation procedure to potentially improve the PLS-POS solution. PLS-POS also requires researchers to specify the search depth, which should be equal to the number of observations. Finally, researchers can choose among different optimization criteria. The weighted R^2 of all endogenous constructs in the PLS path model should generally be preferred as optimization criterion. Key results of the PLS-POS output include the observations' final partition, the segments sizes, and the segment-specific path coefficient estimates.

REVIEW QUESTIONS

1. What is the purpose of a latent class analysis?

2. What is FIMIX-PLS? What are the method's strengths and weaknesses?

3. What is PLS-POS? How does the method work?

4. Why should we jointly use FIMIX-PLS and PLS-POS in a latent class analysis?

5. What is the purpose of an ex post analysis? Why is it important?

CRITICAL THINKING QUESTIONS

1. What is the purpose of the normed entropy criterion? Why is its consideration important when deciding on the number of segments?

2. Discuss the interplay between information criteria, entropy, and relative segment sizes in the decision of how many segments to retain from the data.

3. Under which circumstances would we negate the presence of significant unobserved heterogeneity and continue with the aggregate level analysis?

4. Why is collecting sufficient potential explanatory variables in the research design stage important for successfully running an ex post analysis?

5. Critically discuss the following statement: "Since the results of an ex post analysis will never fully match the latent segment structure generated by the latent class methods, the validity of the results is necessarily limited."

KEY TERMS

Akaike's information criterion (AIC)
Bayesian information criterion (BIC)
Cluster analysis
Consistent AIC (CAIC)
Ex post analysis
Expectation-maximization (EM)
   algorithm
FIMIX-PLS
Finite mixture partial least squares
   (FIMIX-PLS)
Hill climbing
Information criteria
k-means clustering
Label switching
Latent class techniques

Minimum description length 5
   (MDL_5)
Model selection criteria
Modified AIC with factor 3 (AIC_3)
Modified AIC with factor 4 (AIC_4)
Normed entropy statistic (EN)
Observed heterogeneity
Optimization criterion
Prediction errors
Prediction-oriented segmentation in PLS-
   SEM (PLS-POS)
Presegmentation
Response-based segmentation techniques
Search depth
Unobserved heterogeneity

SUGGESTED READINGS

Becker, J.-M., Rai, A., Ringle, C. M., & Völckner, F. (2013). Discovering unobserved heterogeneity in structural equation models to avert validity threats. *MIS Quarterly*, *37*(3), 665–694.

Hahn, C., Johnson, M. D., Herrmann, A., & Huber, F. (2002). Capturing customer heterogeneity using a finite mixture PLS approach. *Schmalenbach Business Review*, *54*(3), 243–269.

Hair, J. F., Sarstedt, M., Matthews, L., & Ringle, C. M. (2016). Identifying and treating unobserved heterogeneity with FIMIX-PLS: Part I–Method. *European Business Review*, *28*(1), 63–76.

Matthews, L., Sarstedt, M., Hair, J. F., & Ringle, C. M. (2016). Identifying and treating unobserved heterogeneity with FIMIX-PLS: Part II—a case study. *European Business Review*, *28*(1), 208–224.

Sarstedt, M., Becker, J.-M., Ringle, C. M., & Schwaiger, M. (2011). Uncovering and treating unobserved heterogeneity with FIMIX-PLS: Which model selection criterion provides an appropriate number of segments? *Schmalenbach Business Review*, *63*(1), 34–62.

Sarstedt, M., Radomir, L., Moisescu, O. I., & Ringle, C. M. (2022). Latent class analysis in PLS-SEM: A review and recommendations for future applications. *Journal of Business Research*, 138, 398–407.

Sarstedt, M., Ringle, C. M., & Hair, J. F. (2017). Treating unobserved heterogeneity in PLS-SEM: A multi-method approach. In R. Noonan & H. Latan (Eds.), *Partial least squares structural equation modeling: Basic concepts, methodological issues and applications* (pp. 197–217). Springer.

GLOSSARY

AIC: see *Akaike's information criterion*.

AIC$_3$: see *Modified AIC with factor 3*.

AIC$_4$: see *Modified AIC with factor 4*.

Akaike's information criterion (AIC): an information criterion that allows assessing the relative goodness-of-fit of a segmentation solution produced by FIMIX-PLS. A smaller AIC value for a certain number of segments indicates a better fit. The AIC has a strong tendency to overestimate the correct number of segments.

Alpha inflation: refers to the fact that the more tests you conduct at a certain significance level, the more likely you are to claim a significant result when this is not so (i.e., a Type I error).

Artifacts: human-made concepts, which are typically measured with formative indicators.

Attribute: an element of the construct definition. It defines the general type of property to which the focal construct refers, such as an attitude.

Average variance extracted (AVE): the degree to which the construct explains the variance of its indicators.

Bandwidth-fidelity tradeoff: a practical dilemma resulting from the tradeoff between using measures that will cover the majority of variation in a trait (domain-level measurement) or measures that will assess a few specific traits (facet-level measurement) more precisely.

Bayesian information criterion (BIC): an information criterion that allows assessing the relative goodness-of-fit of a segmentation solution produced by FIMIX-PLS. A smaller BIC value for a certain number of segments indicates a better fit. The criterion generally performs well for identifying the number of segments in FIMIX-PLS.

Bias-corrected and Bonferroni-adjusted confidence intervals: a confidence interval type used for testing the significance of multiple tetrads considered per measurement model.

BIC: see *Bayesian information criterion (BIC)*.

Bonferroni correction: a method used to counteract the increase in the familywise error rate when performing multiple comparisons across several groups of data.

Bootstrap MGA: compares the bootstrap estimates of a parameter across two groups. By counting the number of occurrences where the bootstrap estimate of the first group is larger than those of the second group, the approach derives a *p* value for a one-tailed test.

Bootstrap samples: make up the number of samples drawn in the bootstrapping procedure. Generally, 5,000 or more samples are recommended.

Bootstrapping: a resampling technique that draws a large number of samples from the original data (with replacement) and estimates models for each sample. It is used to determine standard errors of coefficient estimates to assess the coefficients' statistical significance without relying on distributional assumptions.

Bottleneck table: a tabular representation of the ceiling lines (typically the CR-FDH ceiling line). It shows, row by row, how a certain outcome level of the dependent construct and the corresponding necessary minimum condition level of an independent construct have been achieved.

Bottom-up approach: a way to establish a higher-order construct in which several latent variables (the LOCs) are combined into a single, more abstract construct (the HOC).

CAIC: see *Consistent AIC*.

Causal indicators: an indicator type used in formative measurement models. Constructs measured with causal indicators have an error term, which implies that the indicators do not fully form the construct.

CB-SEM: see *Covariance-based structural equation modeling*.

Ceiling envelopment free disposal hull (CE-FDH) line: given by the scatterplot of the values of the independent (*x*-axis) and dependent construct (*y*-axis). The outermost values of this relationship (lowest *x*-value and highest *y*-value) are associated with a step function that represents the CE-FDH line.

Ceiling regression free disposal hull (CR-FDH) line: given by the scatterplot of the values of the independent (*x*-axis) and dependent construct (*y*-axis). The outermost values of this relationship (lowest *x*-value and highest *y*-value) are associated with a linear regression

function. The CR-FDH line is given by a simple linear regression line through the data points that constitute the *ceiling regression free disposal hull (CE-FDH)* line.

Cluster analysis: a class of methods that partitions a set of objects with the goal of obtaining high similarity within the formed groups and high dissimilarity between the groups.

Clustering: see *Cluster analysis*.

Collect model: a higher-order construct type in which the HOC is a combination of several specific LOCs representing more concrete components that form the general concept. The relationship between HOC and LOCs is formative.

Common factor models: assume that each indicator in a set of observed measures is a linear function of one or more common factors.

Common factor-based SEM: a type of SEM method, which considers the constructs as common factors that explain the covariation between its associated indicators.

Common variance: the variance that an indicator shares with other indicators in the measurement model of a construct.

Composite-based SEM: a type of SEM method, which represents the constructs

as composites, formed by linear combinations of sets of indicator variables.

Composite indicators: an indicator type used in formative measurement models. Constructs measured with composite indicators have no error term, which implies that the indicators fully form the construct.

Composite model approach: an approach to estimating construct proxies. Its objective is to account for the total variance in the observed indicators rather than to explain the correlations between the indicators.

Compositional invariance: exists when the composite scores are equal across groups.

Conceptual variables: broad ideas or thoughts about abstract concepts that researchers establish and propose to measure in their research.

Confidence interval: provides the lower and upper limit of values within which a parameter estimate will fall when repeatedly sampling from the same population. In PLS-SEM, the construction of the interval relies on bootstrapping standard errors.

Configural invariance: exists when constructs are equally parameterized and estimated across groups.

Confirmatory tetrad analysis for PLS-SEM (CTA-PLS): allows statistical testing of whether a measurement model is reflective or formative.

Consistent AIC (CAIC): an information criterion that allows assessing the relative goodness-of-fit of a segmentation solution produced by FIMIX-PLS. A smaller CAIC value for a certain number of segments indicates a better fit. The criterion should be used in conjunction with AIC_3 to determine the number of segments in FIMIX-PLS.

Consistent PLS-SEM (PLSc-SEM): a variant of the standard PLS-SEM algorithm, which provides consistent model estimates that disattenuate the correlations between pairs of latent variables, thereby mimicking CB-SEM results.

Construct definition: the specific way in which a conceptual variable is measured in a particular study, and could differ from one study to another.

Constructs: measure concepts that are abstract, complex, and cannot be directly observed. Constructs are represented in path models as circles or ovals.

Correlation weights: see *Mode A*.

Covariance-based structural equation modeling (CB-SEM): used to confirm (or reject) theories. It does this by determining how well a proposed theoretical model

can estimate the covariance matrix for a sample data set.

CTA-PLS: see *Confirmatory tetrad analysis in PLS-SEM (CTA-PLS)*.

Cubic effect: a nonlinear relationship represented by a polynomial of the degree 3; see also *Nonlinear effect* and *Polynomial*.

Direct effect: the direct relationship between two latent variables in a PLS path model.

Disjoint two-stage approach: uses the LOCs' construct scores estimated in the first stage where the HOC is not included in the PLS path model as indicators of the HOC in the second stage. The approach proves particularly useful for estimating formative (i.e., reflective-formative and formative-formative) higher-order constructs.

Effect indicators: see *Reflective indicators*.

Embedded two-stage approach: uses the LOCs' construct scores estimated in the first stage via the repeated indicators approach as indicators of the HOC in the second stage. The approach proves particularly useful for estimating formative (i.e., reflective-formative and formative-formative) higher-order constructs.

EN: see *Normed entropy statistic*.

Endogenous constructs: serve only as dependent

variables or as both independent and dependent variables in a structural model.

Equality of composite mean values and variances: the final requirement for establishing full measurement invariance.

Equidistant scale: a scale in which the intervals are distributed in equal units. For example, a 5-point Likert scale with two negative categories (completely disagree and disagree), a neutral option, and two positive categories (agree and completely agree) can be considered an equidistant scale.

Error terms: capture the unexplained variance in constructs and indicators when path models are estimated.

Exogenous constructs: serve only as independent variables in a structural model.

Expectation-maximization (EM) algorithm: an iterative algorithm for finding maximum likelihood estimates of parameters in a statistical model. The algorithm alternates between performing an expectation (E) step and a maximization (M) step. The E step creates a function for the expectation of the log-likelihood, which is evaluated using the current estimate of the parameters. The M step computes parameters by maximizing the expected log-likelihood found in the E step. The E and M steps are successively applied until the results stabilize.

Explaining and predicting (EP) theories: a type of theory that involves understanding underlying causes and prediction as well as describing theoretical constructs and the relationships between them.

Explanatory power: provides information about the strength of the assumed causal relationships in a PLS path model.

Ex postanalysis: aims at identifying one or more explanatory variable(s) that match the latent class segmentation results in the best possible way to facilitate a multigroup analysis.

Extended repeated indicators approach: links the antecedent construct of an HOC in a reflective-formative or formative-formative higher-order construct with the LOCs. The effect of the antecedent construct on the HOC is equal to its direct effect plus the sum of all indirect effects via the LOCs.

Factor (score) indeterminacy: means that there is an infinite number of sets of factor scores matching the requirements of a certain common factor model.

Factor weighting scheme: uses the correlations between constructs in the structural model to determine their relationships in the first stage of the PLS-SEM algorithm.

Factor-based SEM: see *Covariance-based structural equation modeling*.

Familywise error rate: the probability of making one or more Type I errors when performing multiple comparisons across several groups of data.

FIMIX-PLS: see *Finite mixture partial least squares (FIMIX-PLS)*.

Finite mixture partial least squares (FIMIX-PLS): a latent class approach that allows for identifying and treating unobserved heterogeneity in PLS path models. The approach applies mixture regressions to simultaneously estimate group-specific parameters and observations' probabilities of segment membership.

Focal object: an element of the construct definition. It refers to the entity to which an attribute is applied.

Forced-choice scale: a scale without a neutral category.

Formative measurement model: a type of measurement model setup in which the direction of the arrows is from the indicator variables to the construct.

Formative-formative higher-order construct: has formative measurement models of all LOCs in the higher-order construct and formative path relationships between the LOCs and the HOC (i.e., the LOCs form the HOC).

Formative-reflective higher-order construct: has formative measurement models

of all LOCs in the higher-order construct and reflective path relationships from the HOC to the LOCs.

Full measurement invariance: confirmed when (1) configural invariance, (2) compositional invariance, and (3) equality of composite mean values and variances are demonstrated.

Hierarchical common factor model: see *Reflective-reflective higher-order construct*.

Hierarchical component model: see *Higher-order construct*.

Higher-order component (HOC): a general construct that represents all underlying LOCs in a higher-order construct.

Higher-order construct: a higher-order structure (usually second-order) that contains several layers of constructs and involves a higher level of abstraction. Higher-order constructs involve a more abstract *higher-order component (HOC)* related to two or more *lower-order components (LOCs)* in a reflective or formative way.

Higher-order model: see *Higher-order construct*.

Hill climbing: a heuristic optimization method that belongs to the family of local search procedures. The method starts with a random partition into a prespecified number of segments and iteratively improves this solution one element at a time until it arrives at a (locally) optimized solution.

HOC: see *Higher-order construct.*

Impact-performance map: a graphical representation of the impact-performance map results; see *Importance-performance map analysis (IPMA).*

Importance: a term used in the context of *IPMA.* It is equivalent to the (unstandardized) total effect of some latent variable on the target variable.

Importance-performance map analysis (IPMA): extends the standard PLS-SEM results reporting of path coefficient estimates by adding a dimension to the analysis that considers the average values of the latent variable scores. More precisely, the IPMA relates structural model total effects on a specific target construct with the average latent variable scores of this construct's predecessors.

Importance-performance matrix: see *Importance-performance map analysis (IPMA).*

Indicators: directly measured observations (raw data), generally referred to as either *items* or *manifest variables*, represented in path models as rectangles.

Indirect effect: represents a relationship between two latent variables via a third (e.g., mediator) construct in the PLS path model. If p_1

is the relationship between a latent variable and the mediator variable, and p_2 is the relationship between the mediator variable and the endogenous latent variable, the indirect effect is the product of path p_1 and path p_2.

Information criteria: statistical measures of the relative quality of a certain segment solution that contrast the fit (i.e., the likelihood) of a solution and the number of parameters used to achieve that fit, which increase with the number of segments. The smaller the value of a certain information criterion, the better the segmentation solution.

Inner model: see *Structural model.*

In-sample predictive power: see *Explanatory power.*

Interaction term: an auxiliary variable entered into the path model to account for the quadratic effect, when considering self-interaction; see also *Self-interaction.*

IPMA: see *Importance-performance map analysis.*

Items: see *Indicators.*

Jangle fallacy: describes the inference that two measures (e.g., scales) with different names measure different constructs.

k-means clustering: a group of nonhierarchical clustering algorithms that work by partitioning observations

into a predefined number of groups and then iteratively reassigning observations until some numeric goal related to cluster distinctiveness is met.

Label switching: occurs when the label of a specific segment changes from one FIMIX-PLS run to the next.

Latent class techniques: a class of approaches that facilitates uncovering unobserved heterogeneity. Different approaches have been proposed, which draw on, for example, finite mixture, genetic algorithm, or hill-climbing approaches to PLS-SEM.

Latent variable: see *Constructs.*

Latent variable scores: the values calculated to represent latent variables.

Linear effect: represented by a straight line when plotted as a graph.

Linear relationship: see *Linear effect.*

LOC: see *Lower-order components.*

Log transformation: a type of data transformation to account for nonlinear relationships, which applies a base 10 logarithm to every observation.

Lower-order components (LOCs): a subdimension of the HOC in a higher-order construct.

Manifest variables: see *Indicators.*

Measurement equivalence: see *Measurement invariance.*

Measurement invariance: deals with the comparability of responses to sets of items across groups. Among other things, measurement invariance implies that the categorical moderator variable's effect is restricted to the path coefficients and does not involve group-related differences in the measurement models.

Measurement invariance of composite models (MICOM) procedure: a series of tests to assess invariance of measures (constructs) across multiple groups of data. The procedure comprises three steps that test different aspects of measurement invariance: (1) configural invariance (i.e., equal parameterization and way of estimation), (2) compositional invariance (i.e., equal indicator weights), and (3) equality of composite mean values and variances.

Measurement model: an element of a path model that contains the indicators and their relationships with the constructs.

Measurement model misspecification: describes the use of a reflective measurement model when it should be formative or the use of a formative measurement model when it should be reflective. Measurement model misspecification usually

yields invalid results and misleading conclusions.

Mediation: represents a situation in which a mediator variable to some extent absorbs the effect of a latent variable on an endogenous latent variable in the PLS path model.

Mediator model: see *Mediation.*

Metric scale: a type of measurement scale that has a constant unit of measurement so that the distance between the scale points is equal (interval scale), thereby allowing for the interpretation of the scale points' absolute differences. In case the scale additionally has an absolute zero point, ratios among the scale points can be interpreted (ratio scale).

Metrological uncertainty: the dispersion of the measurement values that can be attributed to the object or concept being measured.

MICOM: see *Measurement invariance of composite models (MICOM) procedure.*

MIMIC model: see *Multiple indicators and multiple causes model.*

Minimum description length 5 (MDL$_5$): an information criterion that allows assessing the relative goodness-of-fit of a segmentation solution produced by FIMIX-PLS. A smaller MDL$_5$ value for a certain number of segments indicates a better fit. The MDL$_5$ has a strong tendency to underestimate

the correct number of segments.

Mode A: an approach to estimate construct scores from its associated indicators, which uses the bivariate correlations between each indicator and the construct as weights (correlation weights).

Mode B: an approach to estimate construct scores from its associated indicators, which determines the weights by regressing the construct on its indicators (regression weights).

Model selection criteria: see *Information criteria.*

Moderation: occurs when the effect of a latent variable on an endogenous latent variable depends on the values of a third variable, referred to as a moderator variable.

Modified AIC with factor 3 (AIC$_3$): an information criterion that allows assessing the relative goodness-of-fit of a segmentation solution produced by FIMIX-PLS. A smaller AIC$_3$ value for a certain number of segments indicates a better fit. The criterion should be used in conjunction with CAIC to determine the number of segments in FIMIX-PLS.

Modified AIC with factor 4 (AIC$_4$): an information criterion that allows assessing the relative goodness-of-fit of a segmentation solution produced by FIMIX-PLS. A smaller AIC$_4$ value for a certain number of segments indicates a better fit. The criterion generally performs well for identifying the number of segments in FIMIX-PLS.

Multigroup analysis (MGA): tests whether parameters (mostly path coefficients) differ significantly between two groups. Research has proposed a range of approaches to multigroup analysis, which rely on the bootstrapping or permutation procedure.

Multimethod MGA approach: involves applying different MGA.

Multiple battery model: a higher-order construct type in which the LOCs measure the same construct at different points of time using data from the same set of respondents.

Multiple indicators and multiple causes (MIMIC) model: a type of structural equation model that incorporates both formative and reflective indicators to measure latent constructs.

Multiple testing problem: occurs when the Type I error of a series of tests increases exponentially.

Necessary condition analysis (NCA): contrasts dependent and independent variables to identify necessary condition thresholds that have to be met to obtain a certain outcome. This analysis supports researchers in identifying necessary conditions for their outcomes.

Necessity effect size *d*: indicates whether a construct

is necessary for achieving a certain outcome. A necessary condition should have at least a small effect size d (i.e., ≥ 0.1), which is significant.

Necessity logic: implies that an outcome—or a certain level of an outcome—can only be achieved if the necessary condition (e.g., a certain indicator or construct) is in place or has reached a certain level. If the necessary condition is not in place, the outcome will not materialize (i.e., the condition is necessary, but not sufficient for an outcome).

Nominal scale: a measurement scale in which numbers are assigned that can be used to identify and classify objects (e.g., people, companies, products).

Nonlinear effect: is not represented by a straight line when plotted on a graph but by a curve.

Nonlinear relationship: see *Nonlinear effect*.

Nonparametric distance-based test: allows testing whether the complete model is different across two (or more) groups. The test applies the permutation procedure to compare the average (squared Euclidean or geodesic) distance of the model-implied indicator correlation matrix across the groups.

Nonredundant tetrads: tetrads considered for significance testing in CTA-PLS.

Normed entropy statistic (EN): a statistical measure

that uses the observations' probabilities of segment membership as an indication of how well segments are separated. Values above 0.50 permit a clear-cut classification of data into the segments.

Observed heterogeneity: occurs when the sources of heterogeneity are known and can be traced back to observable characteristics such as demographics (e.g., gender, age, income).

Optimization criterion: a certain measure whose value defines the quality of a tested set of parameters.

Ordinal scale: a measurement scale in which the assigned numbers indicate the relative positions of objects in an ordered series.

Orthogonalizing approach: an approach to model the nonlinear (e.g., quadratic) term with orthogonal indicators.

Outer models: see *Measurement model*.

Out-of-sample predictive power: see *Predictive power*.

Parametric MGA: a multigroup variant, representing a modified version of a two-independent-samples *t* test. The test comes in two forms—parametric MGA (equal) or parametric MGA (unequal)—depending on whether the population variances of the groups are considered to be equal or not.

Parametric MGA (equal): see *Parametric MGA*.

Parametric MGA (unequal): see *Parametric MGA*.

Partial least squares algorithm: see *PLS-SEM algorithm*.

Partial least squares path modeling (PLS-SEM): see *Partial least squares structural equation modeling*.

Partial least squares regression (PLS-R): an approach designed to reduce the problem of multicollinearity in regression models. It uses a principal components analysis that extracts linear composites of the independent variables and their respective scores, taking into consideration the relationship between the independent and dependent variables, and maximizing the explanation of the dependent variable.

Partial least squares structural equation modeling (PLS-SEM): a composite-based method to estimate structural equation models. The goal is to maximize the explained variance of the endogenous latent variables and their indicators.

Partial measurement invariance: is confirmed when (1) configural invariance and (2) compositional invariance are demonstrated.

Path weighting scheme: uses the results of partial regression models to determine the relationships between the constructs in the structural model in the first stage of the PLS-SEM algorithm.

Performance: the term used in the context of *IPMA* to represent the mean value of the unstandardized (and rescaled) scores of a latent variable or an indicator.

Permutation: generates a reference distribution of some set of parameters from the actual data by randomly exchanging observations between the groups multiple times.

Permutation MGA: randomly permutes observations between two groups and re-estimates the model to derive a test statistic for the group differences.

PLS algorithm: see *Partial least squares algorithm*.

PLSc: see *Consistent PLS-SEM*.

PLS-POS: see *Prediction-oriented segmentation in PLS-SEM (PLS-POS)*.

PLS-R: see *Partial least squares regression*.

PLS-SEM: see *Partial least squares structural equation modeling*.

PLS-SEM algorithm: the heart of the method. Based on the PLS path model and the indicator data available, the algorithm computes the scores of all latent variables in the model, which in turn serve for estimating all path model relationships.

PLS-SEM bias: refers to PLS-SEM's property that structural model relationships are slightly underestimated and relationships in the measurement models are slightly overestimated compared to CB-SEM when using the method on common factor model data. This difference can be attributed to the methods' different handling of the latent variables in the model estimation but is negligible in most settings typically encountered in empirical research.

Polynomial: a mathematical expression consisting of a sum of terms, whereby each term includes a variable raised to a power and multiplied by a coefficient.

Polynomial degree: determines number of terms that are summed in a polynomial; see also *Polynomial*.

Prediction errors: used in PLS-POS to reassign observations from one segment to the other. They correspond to the observation's squared differences between the predicted and actual scores of the endogenous latent variable(s).

Prediction-oriented segmentation in PLS-SEM (PLS-POS): a distance-based segmentation method for PLS-SEM.

Predictive power: indicates a model's ability to predict new or future observations.

Presegmentation: an option in PLS-POS that, in the first round, assigns all observations at the same time to their closest segment.

Then, in the subsequent iterations, PLS-POS reassigns only one observation per iteration.

Principal components regression: performs a principal components analysis on the independent variables, and the principal components are used as predictive/explanatory variables for the dependent variable. It focuses on reducing the dimensionality of the independent variables without taking into account the relationship between the independent and dependent variables.

Priority map analysis: see *Importance-performance map analysis (IPMA).*

Product indicator approach: an approach to model the nonlinear (e.g., quadratic) term. It involves multiplying all indicators of the exogenous latent variable to establish a measurement model of the nonlinear (e.g., quadratic) term. The approach is only applicable for reflectively measured exogenous latent variables.

Product indicators: indicators of an interaction term, generated by multiplication of each indicator of a construct with each indicator of the moderator variable; see also *Product indicator approach.*

Quadratic effect: represented by a curved nonlinear relationship characterized by a polynomial of the degree 2; see also *Nonlinear effect* and *Polynomial.*

Reflective-formative higher-order construct: has reflective measurement models of all LOCs in the higher-order construct and formative path relationships from the LOCs to the HOC.

Reflective indicators: indicators of a *reflective measurement model.*

Reflective measurement model: a type of measurement model setup in which the direction of the arrows is from the construct to the indicator variables, indicating the assumption that the construct causes the measurement (more precisely, the covariation) of the indicator variables. Indicators in reflective measurement models are also referred to as effect indicators.

Reflective-reflective higher-order construct: has reflective measurement models of all LOCs in the higher-order construct and reflective path relationships from the HOC to the LOCs; also referred to as *Hierarchical common factor model.*

Regression weights: see *Mode B.*

Reliability coefficient ρ_A: a measure of internal consistency reliability.

Repeated indicators approach: a type of measurement model setup in higher-order models that uses the indicators of the LOCs as indicators of the HOC to identify the higher-order construct in a PLS path model.

Rescaling: changes the values of a variable's scale to fit a predefined range (e.g., 0 to 100).

Response-based segmentation techniques: see *Latent class techniques.*

Sampling weights: assign the observations different importance in the parameter estimation in order to obtain unbiased estimates of the population effects.

Search depth: a parameter in PLS-POS that defines the maximum number of observations considered for reassignment to another segment.

Second-order construct: see *Hierarchical component model.*

Self-interaction: occurs when the effect of an exogenous latent variable on an endogenous latent variable depends on the values of the exogenous latent variable.

Single-item measurement: uses only a single item to measure a construct. Since the construct is equal to its measure, the indicator loading is 1.00, making conventional reliability and convergent validity assessments inappropriate.

Spread model: a higher-order construct type in which the HOC is manifested in several more specific LOCs. The relationship between HOC and LOCs is reflective.

Stand-alone higher-order construct: a higher-order

construct that is not embedded in a greater nomological net of constructs.

Structural model: includes the construct and their relationships as derived from theory and logic.

Sufficiency logic: assumes that several factors contribute to the outcome and that one can compensate for another.

Sum scores: represent a naive way to determine the latent variable scores. Instead of estimating the relationships in the measurement models, sum scores use the same weight for each indicator per measurement model (equal weights) to determine the latent variable scores. As such, the sum scores approach does not account for measurement error.

Tetrads (τ): the difference of the product of a pair of covariances and the product of another pair of covariances. In reflective measurement models, this difference is assumed to be zero or at least close to zero; that is, the tetrad is expected to vanish. Nonvanishing tetrads in a latent variable's measurement model cast doubt on its reflective specification, suggesting a formative specification.

Theoretical model: a set of equations with variables that formalize a theory.

Top-down approach: a way to establish a higher-order construct in which a more abstract construct (the HOC) is defined that consists of several components (the LOCs).

Total effect: the sum of the direct effect and the indirect effect(s) between a latent variable and an endogenous latent variable in the PLS path model.

Two-stage approach (nonlinear effects): an approach to model the nonlinear (e.g., quadratic) term. The approach can be used for any kind of exogenous construct no matter whether it is measured reflectively, formatively, or represents a single item construct.

Type I higher-order construct: see *Reflective-reflective higher-order construct*.

Type II higher-order construct: see *Reflective-formative higher-order construct*.

Type III higher-order construct: see *Formative-reflective higher-order construct*.

Type IV higher-order construct: see *Formative-formative higher-order construct*.

Unobserved heterogeneity: occurs when the sources of heterogeneous data structures are not (fully) known.

Vanishing tetrads: see *Tetrads*.

Weighted PLS-SEM: a modified version of the original PLS-SEM algorithm allows the researcher to incorporate sampling weights.

Welch-Satterthwaite *t* test: see *Parametric MGA*.

REFERENCES

Abdi, H. (2010). Partial least squares regression and projection on latent structure regression (PLS regression). *WIREs Computational Statistics*, *2*(1), 97–106.

Agarwal, R., & Karahanna, E. (2000). Time flies when you're having fun: Cognitive absorption and beliefs about information technology usage. *MIS Quarterly*, *24*(4), 665–694.

Aguinis, H., Beaty, J. C., Boik, R. J., & Pierce, C. A. (2005). Effect size and power in assessing moderating effects of categorical variables using multiple regression: A 30-year review. *Journal of Applied Psychology*, *90*(1), 94–107.

Aguirre-Urreta, M. I., & Rönkkö, M. (2018). Statistical inference with PLSc using bootstrap confidence intervals. *MIS Quarterly*, *42*(3), 1001–1020.

Akaike, H. (1973). Information theory and an extension of the maximum likelihood principle. In B. N. Petrov & F. Csaki (Eds.), *Second international symposium on information theory* (pp. 267–281). Academiai Kiadó.

Al-Gahtani, S. S., Hubona, G. S., & Wang, J. (2007). Information technology (IT) in Saudi Arabia: Culture and the acceptance and use of IT. *Information & Management*, *44*(8), 681–691.

Ali, F., Rasoolimanesh, S. M., Sarstedt, M., Ringle, C. M., & Ryu, K. (2018). An assessment of the use of partial least squares structural equation modeling (PLS-SEM) in hospitality research. *International Journal of Contemporary Hospitality Management*, *30*(1), 514–538.

Anderberg, M. R. (1973). *Cluster analysis for applications*. Academic Press.

Antioco, M., Moenaert, R. K., Feinberg, R. A., & Wetzels, M. G. M. (2008). Integrating service and design: The influences of organizational and communication factors on relative product and service characteristics. *Journal of the Academy of Marketing Science*, *36*(4), 501–521.

Antonakis, J., Bendahan, S., Jacquart, P., & Lalive, R. (2010). On making causal claims: A review and recommendations. *Leadership Quarterly*, *21*(6), 1086–1120.

Ahrholdt, D. C., Gudergan, S., & Ringle, C. M. (2019). Enhancing loyalty: When improving consumer satisfaction and delight matters. *Journal of Business Research*, *94*(1), 18–27.

Astrachan, C. B., Patel, V. K., & Wanzenried, G. (2014). A comparative study of CB-SEM and PLS-SEM for theory development in family firm research. *Journal of Family Business Strategy*, *5*(1), 116–128.

Babin, B. J., Hair, J. F., & Boles, J. S. (2008). Publishing research in marketing journals using structural equation modelling. *Journal of Marketing Theory and Practice*, *16*(4), 279–286.

Barroso, C., & Picón, A. (2012). Multi-dimensional analysis of perceived switching costs. *Industrial Marketing Management*, *41*(3), 531–543.

Basco, R., Hair, J. F., Ringle, C. M., & Sarstedt, M. (2021). Advancing family business research through modeling nonlinear relationships: Comparing PLS-SEM and multiple regression. *Journal of Family Business Strategy*, *13*(3), 100457.

Bass, F. M. (1969). A new product growth for model consumer durables. *Management Science*, *15*(5), 215–227.

Bayonne, E., Marin-Garcia, J. A., & Alfalla-Luque,

R. (2020). Partial least squares (PLS) in operations management research: Insights from a systematic literature review. *Journal of Industrial Engineering and Management, 13*(3), 565–597.

Becker, J.-M., Cheah, J. H., Gholamzade, R., Ringle, C. M., & Sarstedt, M. (2023). PLS-SEM's most wanted guidance. *International Journal of Contemporary Hospitality Management, 35*(1), 321–346.

Becker, J.-M., & Ismail, I. R. (2016). Accounting for sampling weights in PLS path modeling: Simulations and empirical examples. *European Management Journal, 34*(6), 606–617.

Becker, J.-M., Klein, K., & Wetzels, M. (2012). Hierarchical latent variable models in PLS-SEM: Guidelines for using reflective-formative type models. *Long Range Planning, 45*(5–6), 359–394.

Becker, J.-M., Rai, A., & Rigdon, E. E. (2013). Predictive validity and formative measurement in structural equation modeling: Embracing practical relevance. In *Proceedings of the 34th International Conference on Information Systems.*

Becker, J.-M., Rai, A., Ringle, C. M., & Völckner, F. (2013). Discovering unobserved heterogeneity in structural equation models to avert validity threats. *MIS Quarterly, 37*(3), 665–694.

Becker, J.-M., Ringle, C. M., & Sarstedt, M. (2018). Estimating moderating effects in PLS-SEM and PLSc-SEM: Interaction term generation*data treatment. *Journal of Applied Structural Equation Modeling, 2*(2), 1–21.

Becker, J.-M., Ringle, C. M., Sarstedt, M., & Völckner, F. (2015). How collinearity affects mixture regression results. *Marketing Letters, 26*(4), 643–659.

Bentler, P. M., & Huang, W. (2014). On components, latent variables, PLS and simple methods: Reactions to Rigdon's rethinking of PLS. *Long Range Planning, 47*(3), 136–145.

Bergh, D. D., Boyd, B. K., Byron, K., Gove, S., & Ketchen, D. J. (2022). What constitutes a methodological contribution? *Journal of Management, 48*(7), 1835–1848.

Bergkvist, L. & Eisend, M. (2021). The dynamic nature of marketing constructs. *Journal of the Academy of Marketing Science, 49*(3), 521–541.

Bergkvist, L., & Langner, T. (2017). Construct measurement in advertising research. *Journal of Advertising, 46*(1), 129–140.

Bergkvist, L., & Langner, T. (2019). Construct heterogeneity and proliferation in advertising research. *International Journal of Advertising, 38*(8), 1286–1302.

Binz Astrachan, C., Patel, V. K., & Wanzenried, G. (2014).

A comparative study of CB-SEM and PLS-SEM for theory development in family firm research. *Journal of Family Business Strategy, 5*(1), 116–128.

Bokrantz, J., & Dul, J. (2023). Building and testing necessity theories in supply chain management. *Journal of Supply Chain Management, 59*(1), 48–65.

Bollen, K. A. (1989). *Structural equations with latent variables.* Wiley.

Bollen, K. A. (2002). Latent variables in psychology and the social sciences. *Annual Review of Psychology, 53,* 605–634.

Bollen, K. A. (2011). Evaluating effect, composite, and causal indicators in structural equation models. *MIS Quarterly, 35*(2), 359–372.

Bollen, K. A., & Bauldry, S. (2011). Three Cs in measurement models: Causal indicators, composite indicators, and covariates. *Psychological Methods, 16*(3), 265–284.

Bollen, K. A., & Davis, W. R. (2009). Causal indicator models: Identification, estimation, and testing. *Structural Equation Modeling: A Multidisciplinary Journal, 16*(3), 498–522.

Bollen, K. A., & Diamantopoulos, A. (2017). In defense of causal-formative indicators: A minority report. *Psychological Methods, 22*(3), 581–596.

Bollen, K. A., & Lennox, R. (1991). Conventional wisdom on measurement:

A structural equation perspective. *Psychological Bulletin, 110*(2), 305–314.

Bollen, K. A., & Ting, K.-F. (1993). Confirmatory tetrad analysis. In P. V. Marsden (Ed.), *Sociological methodology* (pp. 147–175). American Sociological Association.

Bollen, K. A., & Ting, K.-F. (2000). A tetrad test for causal indicators. *Psychological Methods, 5*(1), 3–22.

Borsboom, D., Mellenbergh, G. J., & van Heerden, J. (2003). The theoretical status of latent variables. *Psychological Review, 110*(2), 203–219.

Bozdogan, H. (1987). Model selection and Akaike's Information Criterion (AIC): The general theory and its analytical extensions. *Psychometrika, 52*(3), 345–370.

Bozdogan, H. (1994). Mixture-model cluster analysis using model selection criteria in a new information measure of complexity. In H. Bozdogan (Ed.), *Proceedings of the First US/Japan conference on frontiers of statistical modelling: An information approach* (vol. 2, pp. 69–113). Kluwer Academic.

Carlson, J., Gudergan, S. P., Gelhard, C., & Rahman, M. M. (2019). Customer engagement with brands in social media platforms. *European Journal of Marketing, 53*(9), 1733–1758.

Carrión, G. C., Henseler, J., Ringle, C. M., & Roldán, J. L. (2016).

Prediction-oriented modeling in business research by means of PLS path modeling: Introduction to a JBR special section. *Journal of Business Research, 69*(10), 4545–4551.

Cepeda-Carrión, G., Cegarra-Navarro, J.-G., & Cillo, V. (2019). Tips to use partial least squares structural equation modelling (PLS-SEM) in knowledge management. *Journal of Knowledge Management, 23*(1), 67–89.

Cepeda-Carrión, G., Henseler, J., Ringle, C. M., & Roldán, J. L. (2016). Prediction-oriented modeling in business research by means of PLS path modeling: Introduction to a JBR special section. *Journal of Business Research, 69*(10), 4545–4551.

Cheah, J.-H., Amaro, S., & Roldán, J. L. (2023). Multigroup analysis of more than two groups in PLS-SEM: A review, illustration, and recommendations. *Journal of Business Research, 156*, 113539.

Cheah, J.-H., Roldán, J. L., Ciavolino, E., Ting, H., & Ramayah, T. (2021). Sampling weight adjustments in partial least squares structural equation modeling: Guidelines and illustrations. *Total Quality Management & Business Excellence, 32*(13–14), 1594–1613.

Cheah, J.-H., Sarstedt, M., Ringle, C. M., Ramayah, T., & Ting, H. (2018). Convergent validity assessment of formatively measured

constructs in PLS-SEM. *International Journal of Contemporary Hospitality Management, 30*(11), 3192–3210.

Cheah, J.-H., Ting, H., Ramayah, T., Memon, M.A., Cham, T.-H., & Ciavolino, E. (2019). A comparison of five reflective–formative estimation approaches: Reconsideration and recommendations for tourism research. *Quality & Quantity 53*, 1421–1458.

Chin, W. W. (1994). *PLS Graph* [Computer software]. University of Calgary.

Chin, W. W. (1998). The partial least squares approach to structural equation modeling. In G. A. Marcoulides (Ed.), *Modern methods for business research* (pp. 295–358). Lawrence Erlbaum.

Chin, W. W. (2003). A permutation procedure for multigroup comparison of PLS models. In M. Vilares, M. Tenenhaus, P. Colho, V. Esposito Vinzi, & A. Morineau (Eds.), *PLS and related methods: Proceedings of the International Symposium PLS'03* (pp. 33–43). Decisia.

Chin, W. W. (2010). How to write up and report PLS analyses. In V. Esposito Vinzi, W. W. Chin, J. Henseler, & H. Wang (Eds.), *Handbook of partial least squares: Concepts, methods and applications in marketing and related fields* (pp. 655–690). Springer.

Chin, W. W., Cheah, J-H., Liu, Y., Ting, H., Lim, X.-J., & Cham, T. H. (2020).

Demystifying the role of causal-predictive modeling using partial least squares structural equation modeling in information systems research. *Industrial Management & Data Systems*, *120*(12), 2161–2209.

Chin, W. W., & Dibbern, J. (2010). A permutation based procedure for multi-group PLS analysis: Results of tests of differences on simulated data and a cross cultural analysis of the sourcing of information system services between Germany and the USA. In V. Esposito Vinzi, W. W. Chin, J. Henseler, & H. Wang (Eds.), *Handbook of partial least squares: Concepts, methods and applications in marketing and related fields* (pp. 171–193). Springer.

Chin, W. W., & Newsted, P. R. (1999). Structural equation modeling analysis with small samples using partial least squares. In R. H. Hoyle (Ed.), *Statistical strategies for small sample research* (pp. 307–341). Sage.

Cho, G., Hwang, H., Kim, S., Lee, J., Sarstedt, M., & Ringle, C. M. (2023). A comparative study of the predictive power of component-based approaches to structural equation modeling. *European Journal of Marketing*, *57*(6), 1641–1661.

Cho, G., Sarstedt, M., & Hwang, H. (2022). A comparative evaluation of factor- and component-based structural equation modelling approaches under (in)correct construct representations. *British Journal of Mathematical and Statistical Psychology*, *75*(2), 220–251.

Cho, H. C., & Abe, S. (2012). Is two-tailed testing for directional research hypotheses tests legitimate? *Journal of Business Research*, *66*(9), 1261–1266.

Cliff, N. (1983). Some cautions concerning the application of causal modeling methods. *Multivariate Behavioral Research*, *18*(1), 115–126.

Cohen, J. (1988). *Statistical power analysis for the behavioral sciences* (2nd ed.). Lawrence Erlbaum.

Cook, D. R., & Forzani, L. (2023). On the role of partial least squares in path analysis for the social sciences. *Journal of Business Research*, *167*, 114132.

Cook, R. D., & Forzani, L. (2020). Fundamentals of path analysis in the social sciences. *arXiv*, 2011.06436.

Danks, N. P., Sharma, P. N., & Sarstedt, M. (2020). Model selection uncertainty and multi-model inference in partial least squares structural equation modeling (PLS-SEM). *Journal of Business Research*, *113*, 13–24.

Dessart, L., Aldás-Manzano, J., & Veloutsou, C. (2019). Unveiling heterogeneous engagement-based loyalty in brand communities. *European Journal of Marketing*, *53*(9), 1854–1881.

DeVellis, R. F. (2011). *Scale development*. Sage.

Diamantopoulos, A. (1994). Modelling with LISREL: A guide for the uninitiated. *Journal of Marketing Management*, *10*, 105–136.

Diamantopoulos, A. (2005). The C-OAR-SE procedure for scale development in marketing: A comment. *International Journal of Research in Marketing*, *22*(1), 1–9.

Diamantopoulos, A. (2006). The error term in formative measurement models: Interpretation and modeling implications. *Journal of Modelling in Management*, *1*(1), 7–17.

Diamantopoulos, A. (2011). Incorporating formative measures into covariance-based structural equation models. *MIS Quarterly*, *35*(2), 335–358.

Diamantopoulos, A., & Papadopoulos, N. (2010). Assessing the cross-national invariance of formative measures: Guidelines for international business researchers. *Journal of International Business Studies*, *41*(2), 360–370.

Diamantopoulos, A., & Riefler, P. (2011). Using formative measures in international marketing models: A cautionary tale using consumer animosity as an example. *Advances in*

International Marketing, *22*, 11–30.

Diamantopoulos, A., Riefler, P., & Roth, K. P. (2008). Advancing formative measurement models. *Journal of Business Research*, *61*(12), 1203–1218.

Diamantopoulos, A., Sarstedt, M., Fuchs, C., Kaiser, S., & Wilczynski, P. (2012). Guidelines for choosing between multi-item and single-item scales for construct measurement: A predictive validity perspective. *Journal of the Academy of Marketing Science*, *40*(3), 434–449.

Diamantopoulos, A., & Siguaw, J. A. (2006). Formative vs. reflective indicators in measure development: Does the choice of indicators matter? *British Journal of Management*, *17*(4), 263–282.

Diamantopoulos, A., & Winklhofer, H. M. (2001). Index construction with formative indicators: An alternative to scale development. *Journal of Marketing Research*, *38*(2), 269–277.

Dibbern, J., & Chin, W. W. (2005). Multi-group comparison: Testing a PLS model on the sourcing of application software services across Germany and the USA using a permutation based algorithm. In F. W. Bliemel, A. Eggert, G. Fassott, & J. Henseler (Eds.), *Handbuch PLS-Pfadmodellierung. Methode, Anwendung, Praxisbeispiele* (pp. 135–160). Schäffer-Poeschel.

Dijkstra, T. K. (2010). Latent variables and indices: Herman Wold's basic design and partial least squares. In V. Esposito Vinzi, W. W. Chin, J. Henseler, & H. Wang (Eds.), *Handbook of partial least squares: Concepts, methods and applications in marketing and related fields* (pp. 23–46). Springer.

Dijkstra, T. K. (2014). PLS' Janus face—response to Professor Rigdon's "Rethinking partial least squares modeling: In praise of simple methods." *Long Range Planning*, *47*(3), 146–153.

Dijkstra, T. K., & Henseler, J. (2011). Linear indices in nonlinear structural equation models: Best fitting proper indices and other composites. *Quality & Quantity*, *45*(6), 1505–1518.

Dijkstra, T. K., & Henseler, J. (2015a). Consistent and asymptotically normal PLS estimators for linear structural equations. *Computational Statistics & Data Analysis*, *81*, 10–23.

Dijkstra, T. K., & Henseler, J. (2015b). Consistent partial least squares path modeling. *MIS Quarterly*, *39*(2), 297–316.

Dijkstra, T. K., & Schmermelleh-Engel, K. (2014). Consistent partial least squares for nonlinear structural equation models. *Psychometrika*, *79*(4), 585–604.

Do Valle, P. O., & Assaker, G. (2016). Using partial least squares structural equation modeling in tourism research: A review of past research and recommendations for future applications. *Journal of Travel Research*, *55*(6), 695–708.

Duarte, P., Silva, S. C., Linardi, M. A., & Novais, B. (2022). Understanding the implementation of retail self-service check-out technologies using necessary condition analysis. *International Journal of Retail & Distribution Management*, *50*(13), 140–163.

Dul, J. (2016). Necessary condition analysis (NCA): Logic and methodology of "necessary but not sufficient" causality. *Organizational Research Methods*, *19*(1), 10–52.

Dul, J. (2020). *Conducting Necessary Condition Analysis*. Sage.

Dul, J. (2021). *Advances in Necessary Condition Analysis*. Retrieved from https://bookdown.org/ncabook/advanced_nca2/

Dul, J., Hak, T., Goertz, G., & Voss, C. (2010). Necessary condition hypotheses in operations management. *International Journal of Operations & Production Management*, *30*(11), 1170–1190.

Eberl, M. (2010). An application of PLS in multi-group analysis: The need for differentiated corporate-level marketing in the mobile communications industry. In V.

Esposito Vinzi, W. W. Chin, J. Henseler, & H. Wang (Eds.), *Handbook of partial least squares: Concepts, methods and applications in marketing and related fields* (pp. 487–514). Springer.

Eberl, M., & Schwaiger, M. (2005). Corporate reputation: Disentangling the effects on financial performance. *European Journal of Marketing, 39*(6–7), 838–854.

Edwards, J. R. (2001). Multidimensional constructs in organizational behavior research: An integrative analytical framework. *Organizational Research Methods, 4*(2), 144–192.

Eisenbeiss, M., Cornelißen, M., Backhaus, K., & Hoyer, W. D. (2014). Nonlinear and asymmetric returns on customer satisfaction: Do they vary across situations and consumers? *Journal of the Academy of Marketing Science, 42*(3), 242–263.

Ernst, M. D. (2004). Permutation methods: A basis for exact inference. *Statistical Science, 19*(4), 676–685.

Esposito Vinzi, V., L. Trinchera, & S. Amato. (2010). PLS path modeling: From foundations to recent developments and open issues for model assessment and improvement. In V. Esposito Vinzi, W. W. Chin, J. Henseler, & H. Wang (Eds.), *Handbook of partial least squares: Concepts, methods and applications* (pp. 47–82). Springer.

Esposito Vinzi, V., Trinchera, L., Squillacciotti, S., & Tenenhaus, M. (2008). REBUS-PLS: A response-based procedure for detecting unit segments in PLS path modelling. *Applied Stochastic Models in Business and Industry, 24*(5), 439–458.

Fainshmidt, S., Witt, M. A., Aguilera, R. V., & Verbeke, A. (2020). The contributions of qualitative comparative analysis (QCA) to international business research. *Journal of International Business Studies, 51*(4), 455–466.

Falk, R. F., & Miller, N. B. (1992). *A primer for soft modeling.* University of Akron Press.

Finn, A. (2012). Customer delight: Distinct construct or zone of nonlinear response to customer satisfaction? *Journal of Service Research, 15*(1), 99–110.

Fisher, R. A. (1935). *The design of experiments.* Hafner.

Fordellone, M., & Vichi, M. (2020). Finding groups in structural equation modeling through the partial least squares algorithm. *Computational Statistics & Data Analysis, 147,* 106957.

Fornell, C. G., & Bookstein, F. L. (1982). Two structural equation models: LISREL and PLS applied to consumer exit-voice theory. *Journal of Marketing Research, 19*(4), 440–452.

Fornell, C. G., Johnson, M. D., Anderson, E. W., Cha, J., & Bryant, B. E. (1996). The American Customer Satisfaction Index: Nature, purpose, and findings. *Journal of Marketing, 60*(4), 7–18.

Franke, G., & Sarstedt, M. (2019). Heuristics versus statistics in discriminant validity testing: A comparison of four procedures. *Internet Research, 29*(3), 430–447.

Franke, G., Sarstedt, M., & Danks, N. (2021). Assessing measure congruence in nomological networks. *Journal of Business Research, 130,* 318–334.

Fuchs, C., & Diamantopoulos, A. (2009). Using single-item measures for construct measurement in management research: Conceptual issues and application guidelines. *Die Betriebswirtschaft, 69*(2), 197–212.

Ghasemy, M., Jamil, H., & Gaskin, J. E. (2021). Have your cake and eat it too: PLSe2 = ML + PLS. *Quality & Quantity, 55*(2), 497–541.

Ghasemy, M., Teeroovengadum, V., Becker, J.-M., & Ringle, C. M. (2020). This fast car can move faster: A review of PLS-SEM application in higher education research. *Higher Education, 80*(5), 1121–1152.

Gilliam, D. A., & Voss, K. (2013). A proposed procedure for construct definition in marketing. *European Journal of Marketing, 47*(1–2), 5–26.

Good, P. (2000). *Permutation tests: A practical guide to resampling methods for testing hypotheses.* Springer.

Goodhue, D. L., Lewis, W., & Thompson, R. (2012). Does PLS have advantages for small sample size or non-normal data? *MIS Quarterly, 36*(3), 981–1001.

Gregor, S. (2006). The nature of theory in information systems. *MIS Quarterly, 30*(3), 611–642.

Gudergan, S. P., Ringle, C. M., Wende, S., & Will, A. (2008). Confirmatory tetrad analysis in PLS path modeling. *Journal of Business Research, 61*(12), 1238–1249.

Guenther, P., Guenther, M., Ringle, C. M., Zaefarian, G., & Cartwright, S. (2023). Improving PLS-SEM use for business marketing research. *Industrial Marketing Management, 111*, 127–142.

Guide, V. D. R., & Ketokivi, M. (2015). Notes from the editors: Redefining some methodological criteria for the journal. *Journal of Operations Management, 37*(1), v–viii.

Guttman, L. (1955). The determinacy of factor score matrices with implications for five other basic problems of common-factor theory. *British Journal of Statistical Psychology, 8*(2), 65–81.

Haenlein, M., & Kaplan, A. M. (2004). A beginner's guide to partial least squares analysis. *Understanding Statistics, 3*(4), 283–297.

Hahn, C., Johnson, M. D., Herrmann, A., & Huber, F. (2002). Capturing customer heterogeneity using a finite mixture PLS approach. *Schmalenbach Business Review, 54*(3), 243–269.

Hair, J. F., Black, W. C., Babin, B. J., & Anderson, R. E. (2019). *Multivariate data analysis.* (8th ed.). Cengage Learning.

Hair, J. F., Hollingsworth, C. L., Randolph, A. B., & Chong, A. Y. L. (2017). An updated and expanded assessment of PLS-SEM in information systems research. *Industrial Management & Data Systems, 117*(3), 442–458.

Hair, J. F., Howard, M. C., & Nitzl, C. (2020). Assessing measurement model quality in PLS-SEM using confirmatory composite analysis. *Journal of Business Research, 109*, 101–110.

Hair, J. F., Hult, G. T. M., Ringle, C. M., & Sarstedt, M. (2022). *A primer on partial least squares structural equation modeling (PLS-SEM)* (3rd ed.). Sage.

Hair, J. F., Hult T., Ringle C. M., Sarstedt, M., Danks, N., & Ray, S. (2022). *Partial least squares structural equation modeling (PLS-SEM) using R: A workbook.* Springer.

Hair, J. F., Hult, G. T. M., Ringle, C. M., Sarstedt, M., & Thiele, K. O. (2017). Mirror, mirror on the wall: A comparative evaluation of composite-based structural equation modeling methods. *Journal of the Academy of Marketing Science, 45*(5), 616–632.

Hair, J. F., Matthews, L., Matthews, R., & Sarstedt, M. (2017). PLS-SEM or CB-SEM: Updated guidelines on which method to use. *International Journal of Multivariate Data Analysis, 1*(2), 107–123.

Hair, J. F., Moisescu, O. I., Radomir, L., Ringle, C. M., Sarstedt, M., & Vaithilingam, S. (2021). Executing and interpreting applications of PLS-SEM: Updates for family business researchers. *Journal of Family Business Strategy, 12*(3), 100392.

Hair, J. F., Ringle, C. M., & Sarstedt, M. (2011). PLS-SEM: Indeed a silver bullet. *Journal of Marketing Theory and Practice, 19*(2), 139–151.

Hair, J. F., Ringle, C. M., & Sarstedt, M. (2013). Partial least squares structural equation modeling: Rigorous applications, better results and higher acceptance. *Long Range Planning, 46*(1–2), 1–12.

Hair, J. F., Risher, J. J., Sarstedt, M., & Ringle, C. M. (2019). When to use and how to report the results of PLS-SEM. *European Business Review, 31*(1), 2–24.

Hair, J. F., & Sarstedt, M. (2019). Composites vs. factors: Implications for choosing the right SEM method. *Project*

Management Journal, 50(6), 1–6.

Hair, J. F., & Sarstedt, M. (2021). Explanation plus prediction—the logical focus of project management research. *Project Management Journal, 52*(4), 319–322.

Hair, J. F., Sarstedt, M., Matthews, L., & Ringle, C. M. (2016). Identifying and treating unobserved heterogeneity with FIMIX-PLS: Part I—method. *European Business Review, 28*(1), 63–76.

Hair, J. F., Sarstedt, M., Pieper, T., & Ringle, C. M. (2012). The use of partial least squares structural equation modeling in strategic management research: A review of past practices and recommendations for future applications. *Long Range Planning, 45*(5–6), 320–340.

Hair, J. F., Sarstedt, M., & Ringle, C. M. (2019). Rethinking some of the rethinking of partial least squares. *European Journal of Marketing, 53*(4), 566–584.

Hair, J. F., Sarstedt, M., Ringle, C. M., & Mena, J. A. (2012). An assessment of the use of partial least squares structural equation modeling in marketing research. *Journal of the Academy of Marketing Science, 40*(3), 414–433.

Hauff, S., Guerci, M., Dul, J., & van Rhee, H. (2021). Exploring necessary

conditions in HRM research: Fundamental issues and methodological implications. *Human Resource Management Journal, 31*(1), 18–36.

Hay, D. A., & Morris, D. J. (1991). *Industrial economics and organization: Theory and evidence.* Oxford University Press.

Helm, S., Eggert, A., & Garnefeld, I. (2010). Modelling the impact of corporate reputation on customer satisfaction and loyalty using PLS. In V. Esposito Vinzi, W. W. Chin, J. Henseler, & H. Wang (Eds.), *Handbook of partial least squares: Concepts, methods and applications in marketing and related fields* (pp. 515–534). Springer.

Henseler, J. (2017). Bridging design and behavioral research with variance-based structural equation modeling. *Journal of Advertising, 46*(1), 178–192.

Henseler, J., & Chin, W. W. (2010). A comparison of approaches for the analysis of interaction effects between latent variables using partial least squares path modeling. *Structural Equation Modeling, 17*(1), 82–109.

Henseler, J., Dijkstra, T. K., Sarstedt, M., Ringle, C. M., Diamantopoulos, A., Straub, D. W., Ketchen, David J., Hair, Joseph F., Hult, G. Tomas M., & Calantone, R. J. (2014). Common beliefs and reality about

partial least squares: Comments on Rönkkö & Evermann (2013). *Organizational Research Methods, 17*(2), 182–209.

Henseler, J., Fassott, G., Dijkstra, T. K., & Wilson, B. (2012). Analysing quadratic effects of formative constructs by means of variance-based structural equation modelling. *European Journal of Information Systems, 21*(1), 99–112.

Henseler, J., Hubona, G., & Ray, P. A. (2016). Using PLS path modeling in new technology research: Updated guidelines. *Industrial Management & Data Systems, 116*(1), 2–20.

Henseler, J., Ringle, C. M., & Sarstedt, M. (2012). Using partial least squares path modeling in international advertising research: Basic concepts and recent issues. In S. Okazaki (Ed.), *Handbook of research in international advertising* (pp. 252–276). Edward Elgar.

Henseler, J., Ringle, C. M., & Sarstedt, M. (2015). A new criterion for assessing discriminant validity in variance-based structural equation modeling. *Journal of the Academy of Marketing Science, 43*(1), 115–135.

Henseler, J., Ringle, C. M., & Sarstedt, M. (2016). Testing measurement invariance of composites using partial least squares. *International Marketing Review, 33*(3), 405–431.

Henseler, J., Ringle, C. M., & Sinkovics, R. R. (2009).

The use of partial least squares path modeling in international marketing. *Advances in International Marketing*, *20*, 277–320.

Höck, C., Ringle, C. M., & Sarstedt, M. (2010). Management of multi-purpose stadiums: Importance and performance measurement of service interfaces. *International Journal of Services Technology and Management*, *14*(2–3), 188–207.

Holm, S. (1979). A simple sequentially rejective multiple test procedure. *Scandinavian Journal of Statistics*, *6*(2), 65–70.

Hotelling, H. (1957). The relation of the newer multivariate statistical methods to factor analysis. *British Journal of Statistical Psychology*, *10*(2), 69–79.

Hubona, G. (2015). *PLS-GUI* [Computer software]. Retrieved from http://www.pls-gui.com

Hui, B. S., & Wold, H. (1982). Consistency and consistency at large of partial least squares estimates. In K. G. Jöreskog & H. Wold (Eds.), *Systems under indirect observation, Part II* (pp. 119–130). North Holland.

Hult, G. T. M., Hair, J. F., Proksch, D., Ringle, C. M., Sarstedt, M., & Pinkwart, A. (2018). Addressing endogeneity in marketing applications of partial least squares structural equation modeling. *Journal of International Marketing*, *26*(3), 1–21.

Hult, G. T. M., Ketchen, D. J., Griffith, D. A., Finnegan, C. A., Gonzalez-Padron, T., Harmancioglu, N., Huang, Y., Talay, M. B., & Cavusgil, S. T. (2008). Data equivalence in cross-cultural international business research: Assessment and guidelines. *Journal of International Business Studies*, *39*(6), 1027–1044.

Hwang, H., Sarstedt, M., Cheah, J. H., & Ringle, C. M. (2020). A concept analysis of methodological research on composite-based structural equation modeling: Bridging PLSPM and GSCA. *Behaviormetrika*, *47*(3), 219–241.

Inoue, A., & Kilian, L. (2005). In-sample or out-of-sample tests of predictability: Which one should we use? *Econometric Reviews*, *23*(4), 371–402.

Jarvis, C. B., MacKenzie, S. B., & Podsakoff, P. M. (2003). A critical review of construct indicators and measurement model misspecification in marketing and consumer research. *Journal of Consumer Research*, *30*(2), 199–218.

JCGM/WG1 (2008). Joint Committee for Guides in Metrology/Working Group on the Expression of Uncertainty in Measurement (JCGM/WG1): Evaluation of measurement data-guide to the expression of uncertainty in measurement. Retrieved from https://www.bipm.org/utils/common/documents/jcgm/JCGM_100_2008_E.pd

f (access date: September 9, 2022).

Jedidi, K., Jagpal, H. S., & DeSarbo, W. S. (1997). Finite-mixture structural equation models for response-based segmentation and unobserved heterogeneity. *Marketing Science*, *16*(1), 39–59.

Johnson, R. E., Rosen, C. C., & Chang, C.-H. (2011). To aggregate or not to aggregate: Steps for developing and validating higher-order multidimensional constructs. *Journal of Business Psychology*, *26*(3), 241–248.

Jolliffe, I. T. (1982). A note on the use of principal components in regression. *Journal of the Royal Statistical Society: Series C (Applied Statistics)*, *31*(3), 300–303.

Jones, M. A., Mothersbaugh, D. L., & Beatty, S. E. (2000). Switching barriers and repurchase intentions in services. *Journal of Retailing*, *76*(2), 259–274.

Jöreskog, K. G. (1973). A general method for estimating a linear structural equation system. In A. S. Goldberger & O. D. Duncan (Eds.), *Structural equation models in the social sciences* (pp. 85–112). Academic Press.

Jöreskog, K. G. (1978). Structural analysis of covariance and correlation matrices. *Psychometrika*, *43*(4), 443–477.

Jöreskog, K. G., & Wold, H. (1982). The ML and PLS

techniques for modeling with latent variables: Historical and comparative aspects. In H. Wold & K. G. Jöreskog (Eds.), *Systems under indirect observation, Part I* (pp. 263–270). North-Holland.

Kamakura, W. A. (2015). Measure twice and cut once: The carpenter's rule still applies. *Marketing Letters*, *26*(3), 237–243.

Kaufmann, L., & Gaeckler, J. (2015). A structure review of partial least squares in supply chain management research. *Journal of Purchasing & Supply Management*, *21*(4), 259–272.

Keil, M., Saarinen, T., Tan, B. C. Y., Tuunainen, V., Wassenaar, A., & Wei, K.-K. (2000). A cross-cultural study on escalation of commitment behavior in software projects. *MIS Quarterly*, *24*(2), 299–325.

Kendall, M. G. (1957). *A course in multivariate analysis*. Hafner.

Kenny, D. A. (2015). *Moderation*. Retrieved from http://davidakenny.net/cm/moderation.htm

Kessel, F., Ringle, C. M., & Sarstedt, M. (2010). On the impact of missing values on model selection in FIMIX-PLS. *Proceedings of the 2010 INFORMS Marketing Science Conference*, Cologne, Germany.

Khan, G. F., Sarstedt, M., Shiau, W.-L., Hair, J. F., Ringle, C. M., & Fritze, M. (2019). Methodological research on partial least squares structural equation modeling (PLS-SEM): An analysis based on social network approaches. *Internet Research*, *29*(3), 407–429.

Kiers, H. A. L., & Smilde, A. K. (2007). A comparison of various methods for multivariate regression with highly collinear variables. *Statistical Methods and Applications*, *16*(2), 193–228.

Klesel, M., Schuberth, F., Henseler, J., & Niehaves, B. (2019). A test for multigroup comparison in partial least squares path modeling. *Internet Research*, *29*(3), 464–477.

Klesel, M., Schuberth, F., Niehaves, B., & Henseler, J. (2022). Multigroup analysis in information systems research using PLS-PM: A systematic investigation of approaches. *The Data Base for Advances in Information Systems*, *53*(3), 26–48.

Kline, R. B. (2015). *Principles and practice of structural equation modeling* (4th ed.). Guilford Press.

Kock, N. (2020). *WarpPLS 7.0 user manual*. ScriptWarp Systems.

Kock, N., & Hadaya, P. (2018). Minimum sample size estimation in PLS-SEM: The inverse square root and gamma-exponential methods. *Information Systems Journal*, *28*(1), 227–261.

Kristensen, K., Martensen, A., & Grønholdt, L. (2000). Customer satisfaction measurement at Post Denmark: Results of application of the European customer satisfaction index methodology. *Total Quality Management*, *11*(7), 1007–1015.

Lee, N., & Cadogan, J. W. (2013). Problems with formative and higher-order reflective variables. *Journal of Business Research*, *66*(2), 242–247.

Lee, L., Petter, S., Fayard, D., & Robinson, S. (2011). On the use of partial least squares path modeling in accounting research. *International Journal of Accounting Information Systems*, *12*(4), 305–328.

Li, J. C.-H. (2018). Curvilinear moderation: A more complete examination of moderation effects in behavioral sciences. *Frontiers in Applied Mathematics and Statistics*, *4*(7), 1–11.

Liang, Z., Jaszak, R. J., & Coleman, R. E. (1992). Parameter estimation of finite mixtures using the EM algorithm and information criteria with applications to medical image processing. *IEEE Transactions on Nuclear Science*, *39*(4), 1126–1133.

Liengaard, B., Sharma, P. N., Hult, G. T. M., Jensen, M. B., Sarstedt, M., Hair, J. F., & Ringle, C. M. (2021). Prediction: Coveted, yet forsaken? Introducing a cross-validated predictive ability test in partial least squares path modeling. *Decision Sciences*, *52*(2), 362–392.

Lohmöller, J.-B. (1984). Das Programmsystem LVPLS für Pfadmodelle mit latenten Variablen [The program system LVPLS for path models with latent variables]. *ZA-Information/ Zentralarchiv für Empirische Sozialforschung*, *14*, 44–51.

Lohmöller, J.-B. (1987). *LVPLS 1.8*. Zentralarchiv für Empirische Sozialforschung.

Lohmöller, J.-B. (1989). *Latent variable path modeling with partial least squares*. Springer.

Loo, R. (2002). A caveat on using single-item versus multiple-item scales. *Journal of Managerial Psychology*, *17*(1), 68–75.

MacCallum, R. C., Browne, M. W., & Cai, L. (2007). Factor analysis models as approximations. In R. Cudeck & R. C. MacCallum (Eds.), *Factor analysis at 100: Historical developments and future directions* (pp. 153–175). Lawrence Erlbaum.

MacKenzie, S. B., Podsakoff, P. M., & Podsakoff, N. P. (2011). Construct measurement and validation procedures in MIS and behavioral research: Integrating new and existing techniques. *MIS Quarterly*, *35*(2), 293–334.

Magno, F., Cassia, F., & Ringle, C. M. (2022). A brief review of partial least squares structural equation modeling (PLS-SEM) use in quality management studies. *The TQM Journal*. Advance online publication.

Magno, F., & Dossena, G. (2023). The effects of chatbots' attributes on customer relationships with brands: PLS-SEM and importance–performance map analysis. *The TQM Journal*, *35*(5), 1156–1169.

Manley, S. C., Hair, J. F., Williams, R. I., & McDowell, W. C. (2020). Essential new PLS-SEM analysis methods for your entrepreneurship analytical toolbox. *International Entrepreneurship and Management Journal*, *17*(1), 1805–1825.

Marcoulides, G. A., & Chin, W. W. (2013). You write but others read: Common methodological misunderstandings in PLS and related methods. In H. Abdi, W. W. Chin, V. Esposito Vinzi, G. Russolillo, & L. Trinchera (Eds.), *New perspectives in partial least squares and related methods* (pp. 31–64). Springer.

Marcoulides, G. A., Chin, W. W., & Saunders, C. (2012). When imprecise statistical statements become problematic: A response to Goodhue, Lewis, and Thompson. *MIS Quarterly*, *36*(3), 717–728.

Martens, M., Martens, H., & Wold, S. (1983). Preference of cauliflower related to sensory descriptive variables by partial least squares (PLS) regression. *Journal of the Science of Food and Agriculture*, *34*(7), 715–724.

Masyn, K. E. (2013). Latent class analysis ad finite mixture modeling. In P. Nathan & T. Little (Eds.), *The Oxford handbook of quantitative methods* (pp. 551–611). Oxford University Press.

Mateos-Aparicio, G. (2011). Partial least squares (PLS) methods: Origins, evolution, and application to social sciences. *Communications in Statistics—Theory and Methods*, *40*(13), 2305–2317.

Matthews, L. (2017). Applying multigroup analysis in PLS-SEM: A step-by-step process. In R. Noonan & H. Latan (Eds.), *Partial least squares structural equation modeling: Basic concepts, methodological issues and applications* (pp. 219–243). Springer.

Matthews, L., Sarstedt, M., Hair, J. F., & Ringle, C. M. (2016). Identifying and treating unobserved heterogeneity with FIMIX-PLS: Part II—a case study. *European Business Review*, *28*(1), 208–224.

McCallum, B. T. (1970). Artificial orthogonalization in regression analysis. *The Review of Economics and Statistics*, *52*(1), 110–113.

McDonald, R. P. (1996). Path analysis with composite variables. *Multivariate Behavioral Research*, *31*(2), 239–270.

McLachlan, G. J., & Peel, D. (2000). *Finite mixture models*. Wiley.

McNeish, D., & Wolf, M. G. (2020). Thinking twice about sum scores. *Behavior Research Methods*, *52*(6), 2287–2305.

Memon, M. A., Cheah, J.-H., Ramayah, T., Ting, H., Chuah, F., & Cham, T. H. (2019). Moderation analysis: Issues and guidelines. *Journal of Applied Structural Equation Modeling, 3*(1), i–ix.

Michell, J. (2013). Constructs, inferences and mental measurement. *New Ideas in Psychology, 31*(1), 13–21.

Mikulic, J. (2022). Fallacy of higher-order reflective constructs. *Tourism Management, 89*, 104449.

Mikulic, J., Prebežac, D., & Dabic, M. (2016). Importance-performance analysis: Common misuse of a popular technique. *International Journal of Market Research, 58*(6), 775–778.

Money, K. G., Hillenbrand, C., Henseler, J., & Da Camara, N. (2012). Exploring unanticipated consequences of strategy amongst stakeholder segments: The case of a European revenue service. *Long Range Planning, 45*(5–6), 395–423.

Muthén, B. O. (1989). Latent variable modeling in heterogeneous populations. *Psychometrika, 54*(4), 557–585.

Nguyen, P.-H., Tsai, J.-F., Lin, M.-H., & Hu, Y.-C. (2021). A hybrid model with spherical fuzzy-AHP, PLS-SEM and ANN to predict vaccination intention against COVID-19. *Mathematics, 9*(23), 3075.

Nitzl, C. (2016). The use of partial least squares structural equation modelling (PLS-SEM) in management accounting research: Directions for future theory development. *Journal of Accounting Literature, 37*, 19–35.

Nitzl, C., & Chin, W. W. (2017). The case of partial least squares (PLS) path modeling in managerial accounting. *Journal of Management Control, 28*(2), 137–156.

Noonan, R., & Wold, H. (1982). PLS path modeling with indirectly observed variables: A comparison of alternative estimates for the latent variable. In K. G. Jöreskog & H. Wold (Eds.), *Systems under indirect observations: Part II* (pp. 75–94). North-Holland.

Nunnally, J. C. (1978). *Psychometric theory* (2nd ed.). McGraw Hill.

Nunnally, J. C., & Bernstein, I. (1994). *Psychometric theory.* McGraw-Hill.

Oliver, R. L. (1980). A cognitive model for the antecedents and consequences of satisfaction. *Journal of Marketing Research, 17*(4), 460–469.

Peng, D. X., & Lai, F. (2012). Using partial least squares in operations management research: A practical guideline and summary of past research. *Journal of Operations Management, 30*(6), 467–480.

Petter, S. (2018). "Haters gonna hate": PLS and information systems research. *ACM SIGMIS Database: The DATABASE for Advances in Information Systems, 49*(2), 10–13.

Petter, S., & Hadavi, Y. (2021). With great power comes great responsibility: The use of partial least squares in information systems research. *ACM SIGMIS Database: The DATABASE for Advances in Information Systems, 52*(SI), 10–23.

Petter, S., Straub, D., & Rai, A. (2007). Specifying formative constructs in information systems research. *MIS Quarterly, 31*(4), 623–656.

Polites, G. L., Roberts, N., & Thatcher, J. B. (2012). Conceptualizing models using multidimensional constructs: A review and guidelines for their use. *European Journal of Information Systems, 21*(1), 22–48.

Proschan, M. A., & Brittain E. H. (2020). A primer on strong vs. weak control of familywise error rate. *Statistics in Medicine, 39*(9), 1407–1413.

Rademaker, M. E., Schuberth, F., Schamberger, T., Klesel, M., Dijkstra, T. K., & Henseler, J. (2021). R package cSEM: Composite-based structural equation modeling version 0.4.0 [Computer software]. Retrieved from https://cran.r-project.org/web/packages/cSEM/

Radomir, L., & Moisescu, O. I. (2020). Discriminant validity of the customer-based corporate reputation scale: Some causes for concern. *Journal of Product & Brand Management*, *29*(4), 457–469.

Radomir, L., & Wilson, A. (2018). Corporate reputation: The importance of service quality and relationship investment. In N. K. Avkiran and C. M. Ringle (Eds.), *Partial least squares structural equation modeling: Recent advances in banking and finance*. Springer, 77–123.

Raithel, S., & Schwaiger, M. (2014). The effects of corporate reputation perceptions of the general public on shareholder value. *Strategic Management Journal*, *36*(6), 945–956.

Raithel, S., Wilczynski, P., Schloderer, M. P., & Schwaiger, M. (2010). The value-relevance of corporate reputation during the financial crisis. *Journal of Product and Brand Management*, *19*(6), 389–400.

Ramaswamy, V., DeSarbo, W. S., Reibstein, D. J., & Robinson, W. T. (1993). An empirical pooling approach for estimating marketing mix elasticities with PIMS data. *Marketing Science*, *12*(1), 103–124.

Ramayah, T., Cheah, J.-H., Chuah, F., Ting, H., & Memon, M. A. (2016). *Partial least squares structural equation modeling (PLS-SEM) using SmartPLS 3.0: An updated and practical guide to statistical analysis*. Pearson Malaysia.

Ray, S. Danks, N. P., Valdez, A. C., Velasquez Estrada, J. M., Uanhoro, J., Nakayama, J., Koyan, L., Burbach, L., Bejar, A. H. C., & Adler, S. (2022). R package seminar: Domain-specific language for building and estimating structural equation models version 2.3.0 [Computer software]. Retrieved from h ttps://cran.r-project.org/w eb/packages/seminr/

Reinartz, W., Haenlein, M., & Henseler, J. (2009). An empirical comparison of the efficacy of covariance-based and variance-based SEM. *International Journal of Research in Marketing*, *26*(4), 332–344.

Rhemtulla, M., van Bork, R., & Borsboom, D. (2020). Worse than measurement error: Consequences of inappropriate latent variable measurement models. *Psychological Methods*, *25*(1), 30–45.

Richter, N. F., Cepeda Carrión, G., Roldán, J. L., & Ringle, C. M. (2016). European management research using partial least squares structural equation Modelling (PLS-SEM): Editorial. *European Management Journal*, *34*(6), 589–597.

Richter, N. F., & Hauff, S. (2022). Necessary conditions in international business research: Advancing the field with a new perspective on causality and data analysis. *Journal of World Business*, *57*(5), 101310.

Richter, N. F., Hauff, S., Kolev, A. E., & Schubring, S. (2023). Dataset on an extended technology acceptance model: A combined application of PLS-SEM and NCA. *Data in Brief*, 109190.

Richter, N. F., Hauff, S., Ringle, C. M., & Gudergan, S. P. (2022). The use of partial least squares structural equation modeling and complementary methods in international management research. *Management International Review*, *62*(4), 449–470.

Richter, N. F., Hauff, S., Ringle, C. M., Sarstedt, M., Kolev, A., & Schubring, S. (2023). How to apply necessary condition analysis in PLS-SEM. In J. F. Hair, R. Noonan, & H. Latan (Eds.), *Partial least squares structural equation modeling: Basic concepts, methodological issues and applications* (2nd ed.), forthcoming. Springer.

Richter, N. F., Schlaegel, C., Midgley, D. F., & Tressin, T. (2019). Organizational structure characteristics' influences on international purchasing performance in different purchasing locations. *Journal of Purchasing and Supply Management*, *25*(4), 100523.

Richter, N. F., Schubring, S., Hauff, S., Ringle, C. M., & Sarstedt, M. (2020). When predictors of outcomes are necessary: Guidelines for the combined use of PLS-SEM and NCA. *Industrial Management*

& *Data Systems*, *120*(12), 2243–2267.

Richter, N. F., Sinkovics, R. R., Ringle, C. M., & Schlägel, C. (2016). A critical look at the use of SEM in international business research. *International Marketing Review*, *33*(3), 376–404.

Rigdon, E. E. (2005). Structural equation modeling: Nontraditional alternatives. In B. Everitt & D. Howell (Eds.), *Encyclopedia of statistics in behavioral science* (pp. 1934–1941). Wiley.

Rigdon, E. E. (2012). Rethinking partial least squares path modeling: In praise of simple methods. *Long Range Planning*, *45*(5–6), 341–358.

Rigdon, E. E. (2013). Partial least squares path modeling. In G. R. Hancock & R. O. Mueller (Eds.), *Structural equation modelling: A second course* (2nd ed., pp. 81–116). Information Age Publishing.

Rigdon, E. E. (2014). Rethinking partial least squares path modeling: Breaking chains and forging ahead. *Long Range Planning*, *47*(3), 161–167.

Rigdon, E. E. (2016). Choosing PLS path modeling as analytical method in European management research: A realist perspective. *European Management Journal*, *34*(6), 598–605.

Rigdon, E. E., Becker, J.-M., Rai, A., Ringle, C. M., Diamantopoulos, A.,

Karahanna, E., Straub, D., & Dijkstra, T. (2014). Conflating antecedents and formative indicators: A comment on Aguirre-Urreta and Marakas. *Information Systems Research*, *25*(4), 780–784.

Rigdon, E. E., Becker, J.-M., & Sarstedt, M. (2019). Factor indeterminacy as metrological uncertainty: Implications for advancing psychological measurement. *Multivariate Behavioral Research*, *54*(3), 429–443.

Rigdon, E. E., Preacher, K. J., Lee, N., Howell, R. D., Franke, G. R., & Borsboom, D. (2011). Overcoming measurement dogma: A response to Rossiter. *European Journal of Marketing*, *45*(11), 1589–1600.

Rigdon, E. E., Ringle, C. M., & Sarstedt, M. (2010). Structural modeling of heterogeneous data with partial least squares. In N. K. Malhotra (Ed.), *Review of marketing research* (pp. 255–296). M. E. Sharpe.

Rigdon, E. E., Ringle, C. M., Sarstedt, M., & Gudergan, S. P. (2011). Assessing heterogeneity in customer satisfaction studies: Across industry similarities and within industry differences. *Advances in International Marketing*, *22*, 169–194.

Rigdon, E. E., & Sarstedt, M. (2022). Accounting for uncertainty in the measurement of unobservable marketing phenomena. In H. Baumgartner & B. Weijters (Eds.), *Review of*

Marketing Research, vol. 19 (pp. 53–73). Emerald.

Rigdon, E. E., Sarstedt, M., & Becker, J.-M. (2020). Quantify uncertainty in behavioral research. *Nature Human Behaviour*, *4*(4), 329–331.

Rigdon, E. E., Sarstedt, M., & Ringle, C. M. (2017). On comparing results from CB-SEM and PLS-SEM: Five perspectives and five recommendations. *Marketing ZFP*, *39*(3), 4–16.

Ringle, C. M. (2019). What makes a great textbook? Lessons learned from Joe Hair. In B. J. Babin & M. Sarstedt (Eds.), *The great facilitator: Reflections on the contributions of Joseph F. Hair, Jr. to marketing and business research* (pp. 131–150). Springer.

Ringle, C. M., & Sarstedt, M. (2016). Gain more insight from your PLS-SEM results: The importance-performance map analysis. *Industrial Management & Data Systems*, *116*(9), 1865–1886.

Ringle, C. M., Sarstedt, M., Mitchell, R., & Gudergan, S. P. (2020). Partial least squares structural equation modeling in HRM research. *The International Journal of Human Resource Management*, *31*(12), 1617–1643.

Ringle, C. M., Sarstedt, M., & Mooi, E. A. (2010). Response-based segmentation using finite mixture partial least squares: Theoretical foundations and an application to American customer satisfaction index data. *Annals of Information Systems*, *8*, 19–49.

Ringle, C. M., Sarstedt, M., & Schlittgen, R. (2014). Genetic algorithm segmentation in partial least squares structural equation modeling. *OR Spectrum*, *36*, 251–276.

Ringle, C. M., Sarstedt, M., Schlittgen, R., & Taylor, C. R. (2013). PLS path modeling and evolutionary segmentation. *Journal of Business Research*, *66*(9), 1318–1324.

Ringle, C. M., Sarstedt, M., Sinkovics, N., & Sinkovics, R. R. (2023). A perspective on using partial least squares structural equation modelling in data articles. *Data in Brief*, *48*, 109074.

Ringle, C. M., Sarstedt, M., & Straub, D. W. (2012). A critical look at the use of PLS-SEM in "MIS Quarterly". *MIS Quarterly*, *36*(1), iii–xiv.

Ringle, C. M., Wende, S., & Becker, J.-M. (2015). *SmartPLS 3* [Computer software]. Retrieved from https://www.smartpls.com

Ringle, C. M., Wende, S., & Will, A. (2005). SmartPLS2 [Computer software]. *SmartPLS*. Retrieved from https://www.smartpls.com

Ringle, C. M., Wende, S., & Becker, J.-M. (2022). *SmartPLS 4* [Computer software]. Retrieved from https://www.smartpls.com

Robins, J. (2012). Partial least squares (Editorial). *Long Range Planning*, *45*, 309–311.

Robins, J. (2014). Partial least squares revisited. *Long Range Planning*, *47*(3), 131.

Roldán, J. L., & Sánchez-Franco, M. J. (2012). Variance-based structural equation modeling: Guidelines for using partial least squares in information systems research. In M. Mora, O. Gelman, A. L. Steenkamp, & M. Raisinghani (Eds.), *Research methodologies, innovations and philosophies in software systems engineering and information systems* (pp. 193–221). IGI Global.

Rönkkö, M., & Evermann, J. (2013). A critical examination of common beliefs about partial least squares path modeling. *Organizational Research Methods*, *16*(3), 425–448.

Rönkkö, M., McIntosh, C. N., Antonakis, J., & Edwards, J. R. (2016). Partial least squares path modeling: Time for some serious second thoughts. *Journal of Operations Management*, *47–48*, 9–27.

Rossiter, J. R. (2002). The C-OAR-SE procedure for scale development in marketing. *International Journal of Research in Marketing*, *19*(4), 305–335.

Rossiter, J. R. (2011). *Measurement for the social sciences. The C-OAR-SE method and why it must*

replace psychometrics. Springer.

Rossiter, J. R. (2016). How to use C-OAR-SE to design optimal standard measures. *European Journal of Marketing*, *50*(11), 1924–1941.

Russo, D., & Stol, K.-J. (2021). PLS-SEM for software engineering research: An introduction and survey. *ACM Computing Surveys*, *54*(4), 1–38.

Russo, D., & Stol, K.-J. (2023). Don't Throw the Baby Out With the Bathwater: Comments on "Recent Developments in PLS". *Communications of the Association for Information Systems*, *52*, 700–704.

Salzberger, T., Sarstedt, M., & Diamantopoulos, A. (2016). Measurement in the social sciences: Where C-OAR-SE delivers and where it does not. *European Journal of Marketing*, *50*(11), 1942–1952.

Sarstedt, M. (2008). A review of recent approaches for capturing heterogeneity in partial least squares path modelling. *Journal of Modelling in Management*, *3*(2), 140–161.

Sarstedt, M. (2019). Der Knacks and a silver bullet. In B. J. Babin & M. Sarstedt (Eds.), *The great facilitator: Reflections on the contributions of Joseph F. Hair, Jr. to marketing and business research* (pp. 155–164). Springer.

Sarstedt, M., Becker, J.-M., Ringle, C. M., & Schwaiger,

M. (2011). Uncovering and treating unobserved heterogeneity with FIMIX-PLS: Which model selection criterion provides an appropriate number of segments? *Schmalenbach Business Review, 63*(1), 34–62.

Sarstedt, M., Bengart, P., Shaltoni, A. M., & Lehmann, S. (2018). The use of sampling methods in advertising research: A gap between theory and practice. *International Journal of Advertising, 37*(4), 650–663.

Sarstedt, M., & Cheah, J.-H. (2019). Partial least squares structural equation modeling using SmartPLS: A software review. *Journal of Marketing Analytics, 7*(3), 196–209.

Sarstedt, M., & Danks, N. P. (2022). Prediction in HRM research—a gap between rhetoric and reality. *Human Resource Management Journal, 32*(2), 485–513.

Sarstedt, M., Diamantopoulos, A., & Salzberger, T. (2016). Should we use single items? Better not. *Journal of Business Research, 69*(8), 3199–3203.

Sarstedt, M., Diamantopoulos, A., Salzberger, T., & Baumgartner, P. (2016). Selecting single items to measure doubly-concrete constructs: A cautionary tale. *Journal of Business Research, 69*(8), 3159–3167.

Sarstedt, M., Hair, J. F., Cheah, J.-H., Becker, J.-M., & Ringle, C. M. (2019). How to specify, estimate, and validate higher-order

models. *Australasian Marketing Journal, 27*(3), 197–211.

Sarstedt, M., Hair, J. F., Pick, M., Liengaard, B. D., Radomir, L., & Ringle, C. M. (2022). Progress in partial least squares structural equation modeling use in marketing in the last decade. *Psychology & Marketing, 39*(5), 1035–1064.

Sarstedt, M., Hair, J. F., & Ringle, C. M. (2023). PLS-SEM: Indeed a silver bullet—retrospective observations and recent advances. *Journal of Marketing Theory & Practice, 31*(3), 261–275.

Sarstedt, M., Hair, J. F., Ringle, C. M., Thiele, K. O., & Gudergan, S. P. (2016). Measurement issues with PLS and CB-SEM: Where the bias lies! *Journal of Business Research, 69*(10), 3998–4010.

Sarstedt, M., Henseler, J., & Ringle, C. M. (2011). Multi-group analysis in partial least squares (PLS) path modeling: Alternative methods and empirical results. *Advances in International Marketing, 22,* 195–218.

Sarstedt, M., & Mooi, E. A. (2019). *A concise guide to market research. The process, data, and methods using IBM SPSS statistics* (3rd ed.). Springer.

Sarstedt, M., Radomir, L., Moisescu, O. I., & Ringle, C. M. (2022). Latent class analysis in PLS-SEM: A review and

recommendations for future applications. *Journal of Business Research, 138,* 398–407.

Sarstedt, M., & Ringle, C. M. (2010). Treating unobserved heterogeneity in PLS path modelling: A comparison of FIMIX-PLS with different data analysis strategies. *Journal of Applied Statistics, 37*(8), 1299–1318.

Sarstedt, M., Ringle, C. M., Cheah, J.-H., Ting, H., Moisescu, O. I., & Radomir, L. (2020). Structural model robustness checks in PLS-SEM. *Tourism Economics, 26*(4), 531–554.

Sarstedt, M., Ringle, C. M., & Gudergan, S. P. (2016). Guidelines for treating unobserved heterogeneity in tourism research: A comment on Marques and Reis (2015). *Annals of Tourism Research, 57*(March), 279–284.

Sarstedt, M., Ringle, C. M., & Hair, J. F. (2017). Treating unobserved heterogeneity in PLS-SEM: A multimethod approach. In R. Noonan & H. Latan (Eds.), *Partial least squares structural equation modeling: Basic concepts, methodological issues and applications* (pp. 197–217). Springer.

Sarstedt, M., Ringle, C. M., & Hair, J. F. (2021). Partial least squares. In C. Homburg, M. Klarmann, & A. Vomberg (Eds.), *Handbook of marketing research.* Springer.

Sarstedt, M., Ringle, C. M., Henseler, J., & Hair, J. F. (2014). On the emancipation of PLS-SEM. A commentary on Rigdon (2012). *Long Range Planning*, *47*(3), 154–160.

Sarstedt, M., Ringle, C. M., & Iuklanov, D. (2023). Antecedents and consequences of corporate reputation: A dataset. *Data in Brief*, *48*, 109079.

Sarstedt, M., Ringle, C. M., Smith, D., Reams, R., & Hair, J. F. (2014). Partial least squares structural equation modeling (PLS-SEM): A useful tool for family business researchers. *Journal of Family Business Strategy*, *5*(1), 105–115.

Sarstedt, M., & Schloderer, M. P. (2010). Developing a measurement approach for reputation of non-profit organizations. *International Journal of Nonprofit & Voluntary Sector Marketing*, *15*(3), 276–299.

Sarstedt, M., Schwaiger, M., & Ringle, C. M. (2009). Do we fully understand the critical success factors of customer satisfaction with industrial goods? Extending Festge and Schwaiger's model to account for unobserved heterogeneity. *Journal of Business Market Management*, *3*(3), 185–206.

Sarstedt, M., & Wilczynski, P. (2009). More for less? A comparison of single-item and multi-item measures. *Die Betriebswirtschaft*, *69*(2), 211–227.

Sarstedt, M., Wilczynski, P., & Melewar, T. (2013). Measuring reputation in global markets: A comparison of reputation measures' convergent and criterion validities. *Journal of World Business*, *48*(3), 329–339.

Satterthwaite, F. E. (1946). An approximate distribution of estimates of variance components. *Biometrics Bulletin*, *2*(6), 110–114.

Schlägel, C., & Sarstedt, M. (2016). Assessing the measurement invariance of the four-dimensional cultural intelligence scale across countries: A composite model approach. *European Management Journal*, *34*(6), 633–649.

Schlittgen, R., Ringle, C. M., Sarstedt, M., & Becker, J.-M. (2016). Segmentation of PLS path models by iterative reweighted regressions. *Journal of Business Research*, *69*(10), 4583–4592.

Schlittgen, R., Sarstedt, M., & Ringle, C. M. (2020). Data generation for composite-based structural equation modeling methods. *Advances in Data Analysis and Classification*, *14*(4), 747–757.

Schloderer, M. P., Sarstedt, M., & Ringle, C. M. (2014). The relevance of reputation in the nonprofit sector: The moderating effect of sociodemographic characteristics. *International Journal of Nonprofit and Voluntary Sector Marketing*, *19*(2), 110–126.

Schneeweiß, H. (1991). Models with latent variables: LISREL versus PLS. *Statistica Neerlandica*, *45*(2), 145–157.

Schönemann, P. H., & Wang, M.-M. (1972). Some new results on factor indeterminacy. *Psychometrika*, *37*(1), 61–91.

Schuberth, F., Müller, T., & Henseler, J. (2021). Which equations? An inquiry into the equations in partial least squares structural equation modeling. In I. Kemény & Z. Kun (Eds.), *New perspectives in serving customers, patients, and organizations: A festschrift for Judit Simon* (pp. 96–115). Corvinus University of Budapest.

Schuberth, F., Zaza, S., & Henseler, J. (2023). Partial least squares is an estimator for structural equation models: A comment on Evermann and Rönkkö (2021). *Communications of the Association for Information Systems*, *52*, 711–714.

Schwaiger, M. (2004). Components and parameters of corporate reputation: An empirical study. *Schmalenbach Business Review*, *56*(1), 46–71.

Schwaiger, M., Raithel, S., & Schloderer, M. P. (2009). Recognition or rejection: How a company's reputation influences stakeholder behavior. In J. Klewes & R. Wreschniok (Eds.), *Reputation capital: Building and maintaining trust in the 21st century* (pp. 39–55). Springer.

Schwaiger, M., Sarstedt, M., & Taylor, C. R. (2010). Art for the sake of the corporation: Audi, BMW Group, DaimlerChrysler, Montblanc, Siemens, and Volkswagen help explore the effect of sponsorship on corporate reputations. *Journal of Advertising Research, 50*(1), 77–90.

Schwaiger, M., Witmaier, A., Morath, T., & Hufnagel, G. (2021). Drivers of corporate reputation and its differential impact on customer loyalty. *Marketing ZFP—Journal of Research and Management, 43*(4), 3–27.

Schwarz, G. (1978). Estimating the dimensions of a model. *Annals of Statistics, 6*(2), 461–464.

Şengel, Ü., Genç, G., Işkın, M., Çevrimkaya, M., Assiouras, I., Zengin, B., Sarıışık, M., & Buhalis, D. (2022). The impacts of negative problem orientation on perceived risk and travel intention in the context of COVID-19: A PLS-SEM approach. *Journal of Tourism Futures.* Advance online publication.

Sharma, P. N., Liengaard, B. D., Hair, J. F., Sarstedt, M., & Ringle, C. M. (2023). Predictive model assessment and selection in composite-based modeling using PLS-SEM: Extensions and guidelines for using CVPAT. *European Journal of Marketing, 57*(6), 1662–1677.

Sharma, P. N., Liengaard, B. D., Sarstedt, M., Hair, J.

F., & Ringle, C. M. (2023). Extraordinary claims require extraordinary evidence: A comment on "Recent Developments in PLS". *Communications of the Association for Information Systems, 52,* 739–742.

Sharma, P. N., Sarstedt, M., Shmueli, G., Kim, K., & Thiele, K. (2019). PLS-based model selection: The role of alternative explanations in information systems research. *Journal of the Association for Information Systems, 20*(4), 346–397.

Sharma, P. N., Shmueli, G., Sarstedt, M., Danks, N., & Ray, S. (2018). Prediction-oriented model selection in partial least squares path modeling. *Decision Sciences, 52*(3), 567–607.

Shmueli, G. (2010). To explain or to predict? *Statistical Science, 25*(3), 289–310.

Shmueli, G., Ray, S., Velasquez Estrada, J. M., & Chatla, S. B. (2016). The elephant in the room: Evaluating the predictive performance of PLS models. Journal of Business Research, 69(10), 4552–4564.

Shmueli, G., Sarstedt, M., Hair, J. F., Cheah, J.-H., Ting, H., & Ringle, C. M. (2019). Predictive model assessment in PLS-SEM: Guidelines for using PLSpredict. *European Journal of Marketing, 53*(11), 2322–2347.

Sitar-Taut, D.-A., Mican, D., Frömbling, L., & Sarstedt, M. (2021). Digital socialligators? Social media–induced perceived support during the transition to the COVID-19 lockdown. *Social Science Computer Review,* 08944393211065872.

Spearman, C. (1927). *The abilities of man.* Macmillan.

Squillacciotti, S. (2005). Prediction-oriented classification in PLS path modeling. In T. Aluja, J. Casanovas, V. Esposito Vinzi, & M. Tenenhaus (Eds.), *PLS and marketing: Proceedings of the 4th International Symposium on PLS and related methods* (pp. 499–506). DECISIA.

Squillacciotti, S. (2010). Prediction-oriented classification in PLS path modeling. In V. Esposito Vinzi, W. W. Chin, J. Henseler, & H. Wang (Eds.), *Handbook of partial least squares: Concepts, methods and applications in marketing and related fields* (pp. 219–233). Springer.

Steenkamp, J. B. E. M., & Baumgartner, H. (1998). Assessing measurement invariance in cross national consumer research. *Journal of Consumer Research, 25,* 78–107.

Steiger, J. H. (1979). The relationship between external variables and common factors. *Psychometrika, 44*(1), 93–97.

Steiner, W. J., Siems, F. U., Weber, A., & Guhl, D. (2014). How customer satisfaction with respect to price and quality affects customer retention: An integrated approach considering non-linear effects. *Journal of Business Economics, 84*(6), 879–912.

Steinley, D. (2003). Local optima in k-means clustering: What you don't know may hurt you. *Psychological Methods, 8*(3), 294–304.

Streukens, S., Leroi-Werelds, S., & Willems, K. (2017). Dealing with nonlinearity in importance-performance map analysis (IPMA): An integrative framework in a PLS-SEM context. In H. Latan & R. Noonan (Eds.), *Partial least squares structural equation modeling: Basic concepts, methodological issues and applications* (pp. 367–403). Springer.

Sukhov, A., Olsson, L. E., & Friman, M. (2022). Necessary and sufficient conditions for attractive public transport: Combined use of PLS-SEM and NCA. *Transportation Research Part A: Policy and Practice, 158*, 239–250.

Sultana, T., Dhillon, G., & Oliveira, T. (2023). The effect of fear and situational motivation on online information avoidance: The case of COVID-19. *International Journal of Information Management, 69*, 102596.

Temme, D., & Diamantopoulos, A. (2016). Higher-order models with reflective indicators: A rejoinder to a recent call for their abandonment. *Journal*

of Modelling in Management, 11(1), 180–188.

Tenenhaus, M., Esposito Vinzi, V., Chatelin, Y.-M., & Lauro, C. (2005). PLS path modeling. *Computational Statistics & Data Analysis, 48*(1), 159–205.

Thurstone, L. L. (1947). *Multiple factor analysis*. University of Chicago Press.

Usakli, A., & Kucukergin, K. G. (2018). Using partial least squares structural equation modeling in hospitality and tourism: Do researchers follow practical guidelines? *International Journal of Contemporary Hospitality Management, 30*(11), 3462–3512.

Valencia, J. L., & Diaz-Llanos, F. J. (2003). *Regresión PLS en las ciencias experimentales* [PLS regression in the experimental sciences]. Editorial Complutense.

van der Valk, W., Sumo, R., Dul, J., & Schroeder, R. G. (2016). When are contracts and trust necessary for innovation in buyer-supplier relationships? A necessary condition analysis. *Journal of Purchasing and Supply Management, 22*(4), 266–277.

Vandenberg, R. J., & Lance, C. E. (2000). A review and synthesis of the measurement invariance literature: Suggestions, practices, and recommendations for organizational research. *Organizational Research Methods, 3*(1), 4–70.

Venkatesh, V., Morris, M. G., Davis, G. B., & Davis, F.

D. (2003). User acceptance of information technology: Toward a unified view. *MIS Quarterly, 27*(3), 425–478.

Viswanathan, M. (2022). Measurement error and research design: Some practical issues in conducting research. In H. Baumgartner & B. Weijters (Eds.), *Review of Marketing Research*, vol. 19 (pp. 75–94). Emerald.

Völckner, F., Sattler, H., Hennig-Thurau, T., & Ringle, C. M. (2010). The role of parent brand quality for service brand extension success. *Journal of Service Research, 13*(4), 359–361.

Wanous, J. P., Reichers, A., & Hudy, M. J. (1997). Overall job satisfaction: How good are single-item measures? *Journal of Applied Psychology, 82*(2), 247–252.

Wedel, M., & Kamakura, W. (2000). *Market segmentation. Conceptual and methodological foundations* (2nd ed.). Kluwer Academic.

Welch, B. L. (1947). The generalization of "student's" problem when several different population variances are involved. *Biometrika, 34*(1–2), 28–35.

Wetzels, M., Odekerken-Schroder, G., & van Oppen, C. (2009). Using PLS path modeling for assessing hierarchical construct models: Guidelines and empirical illustration. *MIS Quarterly, 33*(1), 177–195.

Wickens, M. R. (1972). A note on the use of proxy variables. *Econometrica, 40*(4), 759–761.

Wilden, R., & Gudergan, S. (2015). The impact of dynamic capabilities on operational marketing and technological capabilities: Investigating the role of environmental turbulence. *Journal of the Academy of Marketing Science, 43*(2), 181–199.

Willaby, H. W., Costa, D. S. J., Burns, B. D., MacCann, C., & Roberts, R. D. (2015). Testing complex models with small sample sizes: A historical overview and empirical demonstration of what partial least squares (PLS) can offer differential psychology. *Personality and Individual Differences, 84*, 73–78.

Wilson, B. (2010). Using PLS to investigate interaction effects between higher order branding constructs. In V. Esposito Vinzi, W. W. Chin, J. Henseler, & H. Wang (Eds.), *Handbook of partial least squares: Concepts, methods and applications in marketing and related fields* (pp. 621–652). Springer.

Wold, H. (1966). Estimation of principal components and related methods by iterative least squares. In P. R. Krishnaiah (Ed.), *Multivariate analysis* (pp. 391–420). Academic Press.

Wold, H. (1973). Nonlinear iterative partial least squares (NIPALS) modeling: Some current developments. In P. R. Krishnaiah (Ed.), *Multivariate analysis III* (pp. 383–407). Academic Press.

Wold, H. (1980). Model construction and evaluation when theoretical knowledge is scarce: Theory and application of partial least squares. In J. Kmenta & J. B. Ramsey (Eds.), *Evaluation of econometric models* (pp. 47–74). Academic Press.

Wold, H. (1982). Soft modeling: The basic design and some extensions. In K. G. Jöreskog & H. Wold (Eds.), *Systems under indirect observations: Part II* (pp. 1–54). North-Holland.

Wold, H. (1985). Partial least squares. In S. Kotz & N. L. Johnson (Eds.), *Encyclopedia of statistical sciences* (pp. 581–591). Wiley.

Wold, S., Ruhe, A., Wold, H., & Dunn, W. J. (1984). The collinearity problem in linear regression: The partial least squares (PLS) approach to generalized inverses. *SIAM Journal on Scientific and Statistical Computing, 5*(3), 735–743.

Wold, S., Sjöström, M., & Eriksson, L. (2001). PLS-regression: A basic tool of chemometrics. *Chemometrics and Intelligent Laboratory Systems, 58*(2), 109–130.

Yuan, K.-H., & Fang, Y. (2022). Which method delivers greater signal-to-noise ratio: Structural equation modelling or regression analysis with weighted composites? *British Journal of Mathematical and Statistical Psychology.* Advance online publication.

Yuan, K.-H, Y. Wen, & J. Tang. (2020). Regression analysis with latent variables by partial least squares and four other composite scores: Consistency, bias and correction. *Structural Equation Modeling: A Multidisciplinary Journal, 27*(3), 333–350.

Zabukovšek, S. S., Bobek, S., Zabukovšek, U., Kalinić, Z., & Tominc, P. (2022). Enhancing PLS-SEM-enabled research with ANN and IPMA: Research study of enterprise resource planning (ERP) systems' acceptance based on the technology acceptance model (TAM). *Mathematics, 10*(9), 1379.

Zhang, B., Sun, T., Cao, M., & Drasgow, F. (2021). Using bifactor models to examine the predictive validity of hierarchical constructs: Pros, cons, and solutions. *Organizational Research Methods, 24*(3), 530–571.

Zhang, Y., & Schwaiger, M. (2009). Empirical research of corporate reputation in China. *Communicative Business, 1*, 80–104.

Zhou, S., & Wang, Y. (2022). How negative anthropomorphic message framing and nostalgia enhance pro-environmental behaviors during the COVID-19 pandemic in China: An SEM-NCA approach. *Frontiers in Psychology, 13*, 977381.

INDEX

Add group, 153, 193
Advanced modeling and model assessment
nonlinear effects, evaluation of, 63–65
nonlinear relationships, 58–61
PLS-SEM, modeling quadratic effects in, 61–63
results interpretation, 65–66
Aguinis, H., 64
Akaike's information criterion (AIC), 173–174
Alpha inflation, 75–76, 149–150
Artifacts, 7
Attributes of satisfaction, 34
Average variance extracted (AVE), 49, 195

Bandwidth-fidelity tradeoff, 34
Barroso, C., 39
Baumgartner, P., 8
Bayesian information criterion (BIC), 47, 65, 173
Beaty, J. C., 64
Becker, J.-M., 12, 41, 45, 134, 170, 172
Bentler, P. M., 14
Bergkvist, L., 19
Bias-corrected and Bonferroni-adjusted confidence intervals, 76–77, 81
Boik, R. J., 64
Bollen, K. A., 72
Bonferroni correction, 150
alpha inflation, 75–76

Bootstrapping, 67, 145, 194
confidence interval, 50, 78
confirmatory tetrad analysis, 75–76
MGA, 146–148, 160
multigroup analysis, 146–148, 160
nonlinear effects evaluation, 63–65
prediction-oriented segmentation, 136, 167
samples, 146
Bottleneck tables, 95, 102
Bottom-up approach, 36
Bulk change option, 100

Casewise deletion, 184
Categorical moderator variables, 134
Causal indicators, 7
CB-SEM. See Covariance-based SEM; Covariance-based structural equation modeling (CB-SEM)
Ceiling effect size overview, 101
Ceiling envelopment–free disposal hull (CE-FDH) line, 94
Ceiling line effect, 101
Ceiling regression–free disposal hull (CR-FDH) line, 94
Chatelin, Y.-M., 15
Cheah, J.-H., 12, 41, 45
Chemometrics, 2
Chin, W. W., 148
Chi-squared Automatic Interaction Detector (CHAID), 181

Cho, G., 16
Ciavolino, E., 12
Classification and regression trees (CART), 181
Clear groups, 153
Cluster analysis techniques, 135, 168
Collect models, 37
Collinearity, 2, 13, 35, 50, 85, 108, 175
Common factor model approach, 14, 137
CB-SEM versus PLS-SEM estimation differences, 13
measurement invariance assessment, 137, 142
Common variance, 13, 14, 17
Complete data, 119, 153
after casewise deletion, 184, 185, 193
Complete (slower) option, 194
Composite indicators, 7
CB-SEM and, 14
criticism of measurement invariance assessment, 164
partial least squares regression, 2
Composite model approach, 13, 30
Compositional invariance, 137, 139–142
Conceptual variables, 17, 18, 33
operational definitions, 36–37
Confidence interval method, 142, 160

Confidence intervals bias-corrected and Bon-ferroni-adjusted, 50, 81, 143, 156
Configural invariance, 137, 138–139
Confirmatory tetrad analyses (CTA), 80
Confirmatory tetrad analysis for PLS-SEM (CTA-PLS), 72
bias-corrected and Bon-ferroni-adjusted confidence inter-vals, 76–77, 81
bootstrapping, 75
case study illustration, 79–85
cautionary notes
CB-SEM and, 3
formative measurement model specifica-tion, 7
higher-order constructs, 33–41
method, 71–78
motivation for, 33–36
number of indicators, 16, 18, 81
procedure, 48
rules of thumb, 47–48
SmartPLS 3 options, 3
tetrads, 73
Consistent Akaike's Infor-mation Criteria (CAIC), 173–174
Consistent PLS (PLSc) approach, 12–14
Consistent PLS-SEM (PLSc-SEM) algorithm, 12–14
Constructs
formative or reflective, 6
indicators, 6
reliability of measures, 12
single item, 8
Corporate reputation (advanced), 79
Corporate reputation case study

confirmatory tetrad analysis, 78–85
finite mixture partial least squares (FIMIX-PLS), 136, 168
hierarchical component model, 34
importance-performance map analysis, 105
invariance assessment and multigroup analysis, 151
model setup in Smart-PLS, 3
nonlinear effects model, 62
prediction-oriented seg-mentation, 136
Corporate reputation data 336 PLS-POS, 190, 191, 192
Corporate reputation model, 23–25, 28, 49, 52, 53, 70, 79, 121, 152, 184, 193
Corporate reputation (advanced) project, 49, 119, 191, 193
Correlation of indicators, 11
Correlation weights, 10
Counts, 103
percentiles, 103
Covariance-based struc-tural equation modeling (CB-SEM), 3
causal indicators and, 7
composite indicators and, 7, 8
confirmatory tetrad analysis, 74
higher-order constructs, 41
latent class techniques, 136
measurement invariance assessment, 22
model estimation, 12, 19
multiple indicators and multiple causes (MIMIC) model, 13

PLS-SEM comparison, 3, 14, 16, 20
prediction and, 6
specification issues, 7
Create button, 190
Create data file option, 49, 99
Cronbach's alpha, 8, 14, 49, 195
Cross tabs, 181, 192
Cubic effects, 60, 61

Data
group, 194
transformations, 59
view, 99, 191
Delimiter character, 119
Diamantopoulos, A., 8, 72
Dibbern, J., 148
Dimensionality, 2
Direct effect, 112
importance-performance map analysis, 105
Discriminant validity test-ing, 46, 64
Disjoint two-stage approach, 43

Eberl, M., 23
Effect indicators, 6
Effect size, 97, 101–102
f^2 effect size, 64, 70–71
Embedded two-stage approach, 42
Endogenous constructs, 5
Entropy-based measures, 173
Equality of composite mean values and variances, 137, 142–144
Equidistant scale, 108, 119
Error terms, 5–7, 10, 59, 61
Escape character, 119
Esposito Vinzi, V., 15
Example–Corporate repu-tation (advanced) project, 28, 49
Exogenous constructs, 5

Expectation-maximization (EM) algorithm, 171
Expert validity, 139
Explaining and predicting (EP) theories, 20
Explanatory power, 19
Ex post analysis, 181–182, 184, 193, 197
Extended repeated indicators approach, 41

Face validity assessment, 139
Factor-based SEM, 14
Factor (score) indeterminacy, 17
Factor weighting scheme, 10
Familywise error rate, 150
Fang, Y., 19
File name, 190
Finite mixture partial least squares (FIMIX-PLS), 136, 168
case study illustration, 23
combined PLS-POS application, 22, 136
determining number of segments, 169–175, 176
information criteria, 173–175
missing values treatment, 172
rules of thumb, 182
running the procedure, 171–172
simulation studies, 169
Finite mixture (FIMIX) segmentation, 184, 189
Fixed seeds, 80, 187, 189
Focal object, 128
Forced-choice scales, 108
Formative-formative hierarchical component model, 40
Formative-formative higher-order construct, 40

Formative measurement models, 7, 21, 37, 46, 50, 71, 72, 112, 137
F-square, 70
Fuchs, C., 8
Full measurement invariance, 137, 143, 156–157

Generate groups, 153
Genetic algorithms, 136, 168
Gholamzade, R., 41
Gregor, S., 20
Group data sets, 194
Group difference modeling, 145
Group name, 193
Gudergan, S. P., 13, 16, 72, 171

Hair, J. F., 13, 16, 23, 25, 28, 41, 45, 46, 65, 171, 172
Hard modeling, 3
Hauff, S., 91, 93
Henseler, J., 9, 137, 144, 146
Heterogeneity (observed), 167
bootstrap MGA, 146–148
case study illustration, 152–161
composite mean values and variances, 142–144
compositional invariance, 139–142
configural invariance, 138–139
measurement model invariance, 136–138
multigroup analysis (MGA), 144–145
parametric MGA, 145–146
permutation test, 148–149
unobserved and, 134–136

Heterogeneity (unobserved), 167
FIMIX-PLS procedure, 171–172
latent class analysis, 184–197
latent segment structure, 180–182
number of segments, 172–175
observed and, 134–136
PLS path models, 168–171
PLS-POS procedure, 175–180
segment-specific models, 182
Hierarchical common factor models, 37
Hierarchical component models (HCMs), 34
disjoint 2nd stage, 49, 50
redundancy analysis REPU, 50
reflective-formative model, 49, 52, 53
reflective-formative 2nd stage model, 49, 50, 52
Higher-order construct (HOC), 34
case study illustration, 47–53
disjoint two-stage approach, 43–44
embedded two-stage approach, 42
estimating, 44–45
motivation, 33–36
rules of thumb, 47
specifying, 41
terminology, 33–36
types of, 36–41
validating, 45–47
Higher-order model, 34
Hill climbing, 180
HTMT, 46, 194, 196
Huang, W., 14
Hult, G. T. M., 23, 25, 28, 46, 65, 172
Hwang, H., 16

Impact-performance map, 105

Importance performance map analysis (IPMA), 92, 105, 122

Importance-performance matrix, 105

Import data file, 26

Indicator average (IA), 52

Indicator data, 79

Indirect effect, 105

Information criteria, 173

Inner model, 5

In-sample predictive power, 19

Install link, 28

Interaction term, 62

Invert measurement model option, 50

IPMA. *See* Importance performance map analysis (IPMA)

Ismail, I. R., 12

Jangle fallacy, 35

Jarvis, C. B., 72

Jensen, M. B., 65

Jöreskog, K. G., 3

Kaiser, S., 8

Kamakura, W. A., 8

Kenny, D. A., 64

Kim, K., 46

Klein, K., 45

Klesel, M., 144, 172

k-means clustering, 135, 168

Kolev, A., 93

Label Calculate, 49

Label switching, 171

Langner, T., 19

Latent class techniques, 136, 168

Latent variable scores, 49

Lauro, C., 15

Liengaard, B. D., 41, 65

Linear effect, 60

Linear model (LM), 53

Linear relationship, 58

Locale, 119

Log transformation, 59

Lohmöller, J.-B., 9, 26, 37

Lower-order components (LOCs), 34

MacKenzie, S. B., 72

Manifest variable scores, 49

Martens, Harald, 2

Matrix, 121

Matthews, L., 171

Maximum iterations, 184, 189

McNeish, D., 15

Measurement equivalence, 136

Measurement invariance, 136

Measurement invariance of composite models (MICOM) procedure, 133–134, 137–138, 157

Measurement models, 6 misspecification, 71

Metric scale, 108

Metrological uncertainty, 17

Minimum Description Length 5 (MDL$_5$), 173

Missing values, 184 treatment, 119

Mode A, 10

Mode B, 10

Model estimation implications, 19–20

Modeling window, 49, 79, 80, 100, 101, 102, 121, 122

Model selection criteria, 70–71, 173

Moderator analysis, 63

Modified AIC with factor 3 (AIC$_3$), 173

Modified AIC with factor 4 (AIC$_4$), 173

Moisescu, O. I., 182

Most important (faster), 50

Multigroup analysis (MGA), 144

Multimethod Multigroup analysis (MGA) approach, 151

Multiple battery model, 39

Multiple testing problem, 75, 149

Necessary condition analysis (NCA), 92, 101 case study illustration, 99–105 definition, 92 identifying, 94–97 LV scores unstandardized (original scale), 99, 100 method, 92–94 necessity logic, 93 PLS-SEM, 100 purpose, 92–94 results interpretation, 98–99 sufficiency logic, 92

Necessity effect size d, 97

Necessity logic, 93

New project, 26

Niehaves, B., 144

Nominal scale, 108

Nonlinear effect, 59

Nonlinear iterative partial least squares (NIPALS) algorithm, 2

Nonlinear model, 70

Nonlinear relationships, 58 case study illustration, 66–71 evaluation of, 63–65 PLS-SEM, modeling quadratic effects in, 61–63 results interpretation, 65–66

Nonparametric distance-based test, 144

Nonredundant tetrads, 75

Normed entropy statistic (EN), 173

Number of repetitions, 184

Number of segments, 186–187

Number of steps for bottleneck tables option, 101

Nunnally, J. C., 12

Objective criterion, 190

Observed heterogeneity, 135, 167

Oliver, R. L., 20

Open report, 49, 80, 101, 121, 189, 194

Optimization criterion, 178, 189

Ordinal scale, 108

Original difference, 159

Original effect size, 101

Original PLS-SEM algorithm, 9–11

Original Sample (O), 80

Orthogonalizing approach, 62

Outer models, 6

Outer weights, 121

Parameter estimation accuracy, 15–19

Parametric MGA, 145–146

Parametric test, 161

Partial least squares path modeling, 2

Partial least squares regression (PLS-R), 2

Partial least squares structural equation modeling (PLS-SEM), 2

algorithm, 9–11, 49, 50, 70, 79, 121, 194

bias, 16

case study illustration, 23–29, 99–105

consistent PLS-SEM (PLSc-SEM), 12–14

corporate reputation model, 23–25

evolution of, 1–5

measurement, philosophy of, 14–15

model estimation, 9–14, 19–20

model specification, 5–8

organization, 20–22

original PLS-SEM algorithm, 9–11

origins, 1–5

parameter estimation accuracy, 15–19

philosophy of measurement, 14–15

principles of, 14–20

software, 26

weighted PLS-SEM (WPLS) algorithm, 11–12

Partial measurement invariance, 137

Path coefficients, 157, 161, 188

Path weighting scheme, 10

Percentiles, 103

bootstrap, 50, 67

Permutation

mean difference, 159

MGA, 148–149

p values, 101, 102, 155, 156, 159

test, 148–149

Pick, M., 41

Picón, A., 39

Pierce, C. A., 64

PLS path modeling, 3

PLS-SEM. *See* Partial least squares structural equation modeling (PLS-SEM)

Podsakoff, P. M., 72

Polynomial degree, 59

Polynomial representation, 59

Prediction errors, 178

Prediction-oriented segmentation (POS), 188

Prediction-oriented segmentation in PLS-SEM (PLS-POS), 136, 169

Pre-segmentation, 176, 189

Principal components regression, 2

Priority map analysis, 105

Product indicator approach, 62

Quadratic effect, 59, 67

Quality criteria, 70, 125, 126, 188

Radomir, L., 41, 182

Rai, A., 170

Ramayah, T., 12

Random assignment, 189

Random number, 184

generator, 154, 189

Random seeds, 187

Reflective-formative higher-order construct, 39

Reflective indicators, 6

Reflective measurement model, 6, 71

Reflective-reflective higher-order construct (Type I higher-order construct), 37

Regression, 100

weights, 10

Reliability coefficient p_A, 12

Repeated indicators approach, 41

Rescaling, 107

Response-based segmentation techniques, 168

Results report, 49, 50, 70

Richter, N. F., 91, 93

Ringle, C. M., 9, 13, 16, 23, 25, 28, 41, 45, 46, 65, 72, 91, 93, 134, 137, 146, 170, 171, 172, 182

Roldán, J. L., 12

R-square, 188, 190

Salzberger, T., 8

Sample Projects, 28

Sampling weights, 11

Sarstedt, M., 8, 9, 13, 16, 23, 25, 28, 41, 45, 46, 65, 91, 93, 137, 171, 172, 182
Schuberth, F., 144
Schubring, S., 91, 93
Search depth, 177, 189
Second-order construct, 34
Segment sizes, 187, 190
Select columns, 190
Select dataset, 49
Sharma, P. N., 46, 65
Shmueli, G., 46
Significance Level, 80
Single-item measurement, 8
Single-item reliability, 8
Sinkovics, R. R., 146
SmartPLS, 3, 21, 26–29, 52, 99–100, 100, 184, 193–194
Social sciences scholars, 8
Soft modeling basic design, 3
Spread model, 37
Stand-alone higher-order construct, 40
Start calculation, 49, 101, 194, 195
Step 3a (mean), 156
Step 3b (variance), 156
Straub, D. W., 41
Structural model, 5
Sufficiency logic, 92

Sufficiency logic complements, 93
Sum of all constructs weighted R-squares, 189
Sum scores, 15
Synchronize navigation, 52, 70

1-tailed (Prepaid plan *vs.* Contract plan) p value, 161
2-tailed (Prepaid plan *vs.* Contract plan) p value, 161
Tang, J., 14
Target construct, 123
Tenenhaus, M., 15
Tetrads (τ), 73
  nonredundant, 76, 80–81, 85
  vanishing, 74, 76
Thiele, K. O., 13, 16, 46
Ting, H., 12
Ting, K. -F., 72
Top-down approach, 36
Total effect, 105
Two-stage approach (non-linear effects), 62
Two tailed testing, 50
Type I higher-order construct, 37
Type II higher-order construct, 39

Type III higher-order construct, 39
Type IV higher-order construct, 40

Unobserved heterogeneity, 135, 167
Unstandardized, 121

Values, 103
Vanishing tetrads, 74
Völckner, F., 134, 170, 172

Web of Science database, 3
Weighted PLS-SEM (WPLS) algorithm, 11–12
Welch-Satterthwaite t test, 145
Wende, S., 72
Wen, Y., 14
Wetzels, M., 45
Wilczynski, P., 8
Will, A., 72
Wold, H., 2, 3
Wold, Svante, 2
Wolf, M. G., 15
Workspace, 28, 49, 70, 79, 119, 184, 194

Yuan, K.-H., 14, 19